NON-DESTRUCTIVE TESTING

NON-DESTRUCTIVE TESTING

BARRY HULL

B.Eng., M.Met., Ph.D., C.Eng., MIM
Senior Lecturer
Dept. of Metallurgy and Materials Engineering
Sheffield City Polytechnic

VERNON JOHN

B.Sc., M.Sc., C.Eng., MIM, MIMM
Visiting Lecturer
City University
formerly *Senior Lecturer*
School of Mechanical Engineering
The Polytechnic of Central London

MACMILLAN
EDUCATION

First published 1988

Published by
MACMILLAN EDUCATION LTD
Houndmills, Basingstoke, Hampshire RG21 2XS
and London
Companies and representatives
throughout the world

Printed in Hong Kong

British Library Cataloguing in Publication Data
Hull, Barry
    Non-destructive testing.
    1. Non-destructive testing.
    I. Title   II. John, Vernon
    620. 1'127   TA417.2

    ISBN 0-333-46561-X
    ISBN 0-333-35788-4 Pbk

Contents

Plates showing radiographs are between pages 120 and 121

Preface

It is of great importance that both individual components and complete engineering assemblies and structures are free from damaging defects and other possible causes of premature failure. A whole series of inspection instruments and techniques has been evolved over the years and new methods are still being developed to assist in the process of assessing the integrity and reliability of parts and assemblies. Non-destructive testing and evaluation methods are widely used in industry for checking the quality of production, and also as part of routine inspection and maintenance in service.

Despite the obvious importance of the subject, and the fact that most of the inspection methods are based on well-established scientific principles, there is a dearth of publications suitable for use as texts in our universities and colleges. The whole area of non-destructive testing receives scant attention in many engineering degree and diploma courses in the UK and this may be a consequence of a shortage of student texts. The authors, in producing this basic text, hope that it will prove useful to students on engineering courses and, possibly, act as a stimulus for the more widespread introduction of the subject into curricula.

It is also hoped that the work will have a wider circulation than merely academic circles and that it will prove useful to many people in industry. The book is not intended for the industrial practitioners in NDT as the coverage of topics is too general and at too shallow a depth for such individuals. On the other hand we hope that it will be widely read by all personnel concerned with production, general management and marketing who have an interest in quality control but who are non-specialists in non-destructive testing.

Barry Hull
Vernon John

London, 1987

Acknowledgements

We wish to acknowledge the help and assistance given us in the preparation of this work and to thank the various organisations that provided the photographs and radiographs which illustrate the text. We would like to thank especially Chris Brook of Wells–Krautkramer Ltd, Frank Farrow of Pantak Ltd, Roger Heasman of Inspection Instruments Ltd, and Colin Moore of the Kodak Marketing and Education Centre, Hemel Hempstead.
Cover illustration courtesy of Kodak Ltd.

1

Introduction

1.1 Need for inspection

Engineers are well used to assessing the properties of a material by means of standardised tests on prepared test pieces. Much valuable information is obtained from these tests including data on the tensile, compressive, shear and impact properties of the material, but such tests are of a destructive nature. In addition, the material properties, as determined in a standard test to destruction, do not necessarily give a clear guide to the performance characteristics of a complex-shaped component which forms part of some larger engineering assembly.

Defects of many types and sizes may be introduced to a material or a component during manufacture and the exact nature and size of any defects will influence the subsequent performance of the component. Other defects, such as fatigue cracks or corrosion cracks, may be generated within a material during service. (The origins of defects in materials and components are shown in figure 1.1.) It is therefore necessary to have reliable means for detecting the presence of defects at the manufacturing stage and also for detecting and monitoring the rate of growth of defects during the service life of a component or assembly.

Often the first stage in the examination of a component is visual inspection. Examination by the naked eye will not reveal much other than relatively large defects which break through the surface. The effectiveness of visual inspection can be increased through the use of a microscope. The most suitable type of microscope for visual examination of a surface is a stereo microscope. A high degree of magnification is not necessary and the majority of microscopes available for this type of inspection have magnifications in the range from ×5 to ×75. Visual inspection need not be confined to external surfaces. Optical inspection probes, both rigid and flexible, have been developed for the inspection of internal surfaces and these probes may be inserted into cavities, pipes and ducts. Inspection probe systems are discussed in chapter 7.

1

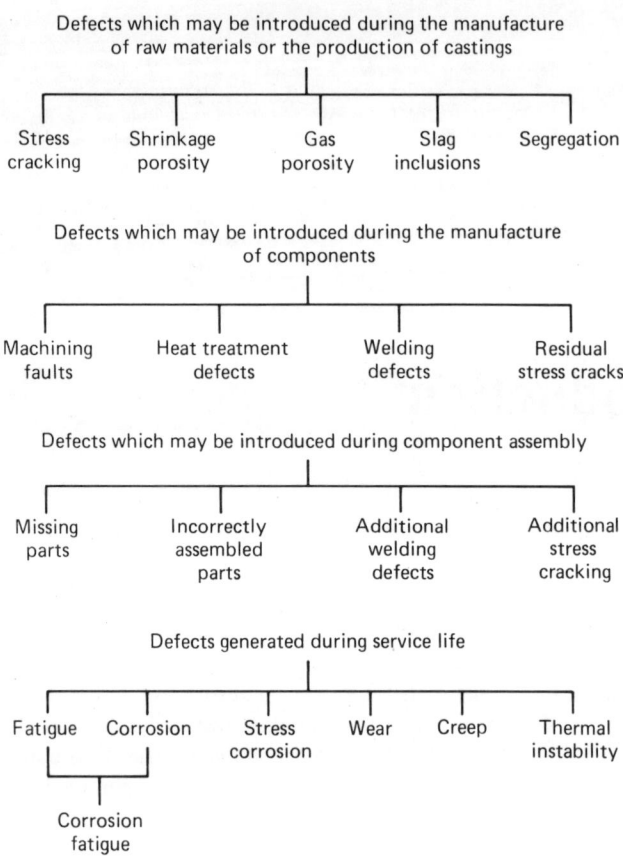

FIGURE 1.1   Origins of some defects found in materials and components.

Using well-established physical principles, a number of non-visual inspection systems have been developed which will provide information on the quality of a material or component and which do not alter or damage the components or assemblies which are tested. The basic principles and major features of the main non-destructive testing (NDT) systems are given in table 1.1.

All these NDT systems co-exist and, depending on the application, may either be used singly or in conjunction with one another. There is some overlap between the various test methods but they are complementary to one another. The fact that, for example, ultrasonic testing can reveal both internal and surface flaws does not necessarily mean that it will be the best method for all inspection applications. Much will depend upon the type of flaw present and the shape and size of the components to be examined.

Table 1.1

System	Features	Applicability
Liquid penetrant	Detection of defects which break the surface	Can be used for any metal, many plastics, glass and glazed ceramics
Magnetic particle	Detection of defects which break the surface and sub-surface defects close to the surface	Can only be used for ferro-magnetic materials (most steels and irons)
Electrical methods (Eddy currents)	Detection of surface defects and some sub-surface defects. Can also be used to measure the thickness of a non-conductive coating, such as paint, on a metal	Can be used for any metal
Ultrasonic testing	Detection of internal defects but can also detect surface flaws	Can be used for most materials
Radiography	Detection of internal defects, surface defects and the correctness of part assemblies	Can be used for many materials but there are limitations on the maximum material thickness

1.2 Types of inspection system

The various non-destructive test methods can be used, in practice, in many different ways and the range of equipment available is extensive. For any one test principle — for example, the use of eddy current techniques — it is possible to purchase a small portable but versatile unit and a selection of test probes at a cost of a few thousand pounds. A skilled operator could use such a unit to detect defects of many types in a wide range of components and materials. At the other end of the scale, a company could make a major capital investment to obtain a purpose-designed fully-automated system which can be incorporated into a production line for the routine inspection of large quantities of rolled metal bar stock. The two pieces of equipment would be vastly different in design, complexity and cost, but both make use of the same physical principles for the detection of abnormalities.

This applies to all the test methods described in this volume. Compact and portable equipment is available which can be used, both inside a test house or out on site, or the basic test principle can be incorporated in some large inspection system dedicated to the examination of large quantities of a single product or a small range of products.

1.3 Quality of inspection

When non-destructive testing systems are used, care must be taken and the processes controlled so that not only qualitative but quantitative information is received and that this information is both accurate and useful. If non-destructive testing is mis-applied it can lead to serious errors of judgement of component quality.

It is necessary that the most dangerous possible failure modes of a component be anticipated, and from this the types and limiting sizes of potentially dangerous defects deduced. In the first instance this is the responsibility of the product designer, and thus it is he who should specify initially what defects are unacceptable and give guidance on the appropriate method of inspection.

It is not always necessary to use an inspection method capable of distinguishing very small discontinuities. For example, in a grey cast iron every graphite flake is a discontinuity. A discontinuity of the same size as a typical graphite flake may be of considerable importance in, say an aluminium forging, and so a high sensitivity test method would be used. If, however, a method of the same sensitivity were used in conjunction with an iron casting, most of the flakes of graphite would be indicated and this mass of information would mask the detection of the larger, but unacceptable, flaws. From this it follows that for the successful application of non-destructive testing the test system and procedures must be suited to the inspection objectives and the types of flaws to be detected, the operator must have sufficient training and experience, and the acceptance standards must be appropriate in defining any undesirable characteristics of a non-conforming part.

If any of these prerequisites is not met, there is a potential for error in detecting and characterising flaws. This is of particular concern if it means a failure to detect flaws that will seriously impair service performance. With inadequate standards, flaws having little or no bearing on product performance may be deemed serious, or significant flaws may be deemed unimportant.

1.4 Reliability of defect detection

In conventional design, a 'design stress' is established by dividing a specific value of yield or proof stress by a suitable safety factor, and this 'strength' is assumed to be representative of the material used to make a component. The fracture mechanics approach to design, however, recognises that flaws can exist in a component, before and during life in service, and attempts to describe quantitatively the effects of such flaws on component integrity. Fracture mechanics describes the capacity of critical structural components to resist the onset of rapid crack growth. Components are characterised by a material property called the *critical stress intensity factor* or *fracture toughness* and the largest flaw that can be tolerated in any specific section of a component. In addition, service environment is taken into account in this quantitative assessment.

The reliability of any non-destructive testing technique is a measure of the efficiency of the technique in detecting flaws of a specific type, shape and size. After inspection has been completed, it can be stated that there is a certain probability that a component is free of defects of a specific type, and shape and size. The higher the level of this probability the greater will be the reliability of the applied technique. However, it must always be borne in mind that non-destructive inspection is carried out for the most part by human beings. Inherently, no two people will perform the same repetitive task in an identical manner all the time. Hence, this additional uncertainty must be accounted for in the evaluation of reliability of inspection, and the worth of accept/reject decisions must be estimated from statistical date.

The role of non-destructive inspection is to guarantee with a level of confidence that cracks corresponding to a critical size for fracture, at the design load, are absent from a component when the component is used in service. It might be necessary to guarantee, with confidence, that cracks smaller than the critical size are absent also. It is important to allow for sub-critical crack growth, especially in components subjected to fatigue loading or to corrosive environments, so that such components can achieve a minimum specified service life before catastrophic failure occurs. In some situations, periodic service inspection or constant monitoring might be necessary to ensure that cracks do not reach a critical size. The use of fracture mechanics concepts in design places a premium on the ability of the various non-destructive methods to detect small cracks. The difference between the critical size and the smallest detectable size becomes the level of safety.

In any particular inspection programme, a significant number of defect indications do not correlate with real flaws. Hence, the probability of identifying a component with no defects of appreciable size is reduced. However, when critical components are involved it is necessary to try and find as many defects as possible, and the tendency to accept all possible defect indications is high, because it is considered better to accept false rejection than take a chance on catastrophic component failure in service. Obviously, an engineer using fracture mechanics concepts is interested in what is the largest defect size that might be overlooked during inspection. Choice of inspection method is dictated, solely, with this as the primary consideration. All other parameters are secondary. For example, ultrasonic inspection of steel components for fatigue cracks, which is relatively easy to carry out, would be rejected in favour of eddy current analysis if the criterion was the detection of cracks of the order of 1.5 mm in length, because the probabilities of detection are 50 per cent and 80 per cent, respectively.

1.5 Benefits of non-destructive test examination

One obvious and clear benefit which can be derived from the judicious use of non-destructive testing is the identification of defects which, if they remained undetected, could result in a catastrophic failure which would be very costly in money and possibly in lives. But the use of these test methods can bring benefits in many ways.

The introduction of any inspection system incurs cost but very often the effective use of suitable inspection techniques will give rise to very considerable financial savings. Not only the type of inspection but also the stages at which inspection is employed is important. It could be very wasteful to reserve the use of a non-destructive test technique for the inspection of small castings or forgings until after all the machining operations have been carried out on the parts. In this case it would be preferable to examine the product before costly machining is commenced and those components with unacceptable flaws rejected. It should be emphasised that not all flaws which may be located at this stage warrant rejection. Some surface discontinuities might be of such a size that they would be removed at the machining stage.

While effective quality control inspection can result in financial savings and help to prevent catastrophic failures in service, it is also true to say that the imposition of too many or too sensitive inspection systems can be very wasteful in terms of both time and money. Excessive inspection may not result in an increase in product performance or reliability. Absolute perfection in a product is impossible to achieve and attempting to get very close to the ideal can prove to be very expensive.

The main non-destructive testing systems are described and discussed in the following chapters.

2

Liquid Penetrant Inspection

2.1 Introduction

Liquid penetrant inspection is a technique which can be used to detect defects in a wide range of components, provided that the defect breaks the surface of the material. The principle of the technique is that a liquid is drawn by capillary attraction into the defect and, after subsequent development, any surface-breaking defects may be rendered visible to the human eye. In order to achieve good defect visibility, the penetrating liquid will either be coloured with a bright and persistent dye or else contain a fluorescent compound. In the former type the dye is generally red and the developed surface can be viewed in natural or artificial light, but in the latter case the component must be viewed under ultra-violet light if indications of defects are to be seen.

The exact origins of this technique are unknown. However, one of the earliest forms of penetrant inspection was the use of carbon black on glazed pottery, to detect glazing cracks. The carbon black was held by the cracks, and hence, their outline was readily visible. Eventually, this was turned to good use as a method of decoration.

Nowadays, liquid penetrant inspection is an important industrial method and it can be used to indicate the presence of defects such as cracks, laminations, laps and zones of surface porosity in a wide variety of components. The method is applicable to almost any component, whether it be large or small, of simple or complex configuration, and it is employed for the inspection of wrought and cast products in both ferrous and non-ferrous metals and alloys, ceramics, glassware and some polymer components.

2.2 Principles of penetrant inspection

There are five essential steps in the penetrant inspection method. These are:

(a) Surface preparation.
(b) Application of penetrant.
(c) Removal of excess penetrant.
(d) Development.
(e) Observation and inspection.

Surface preparation

All surfaces of a component must be thoroughly cleaned and completely dried before it is subjected to inspection. It is important that any surfaces to be examined for defects must be free from oil, water, grease or other contaminants if successful indication of defects is to be achieved.

Application of penetrant

After surface preparation, liquid penetrant is applied in a suitable manner, so as to form a film of penetrant over the component surface. The liquid film should remain on the surface for a period sufficient to allow for full penetration into surface defects.

Removal of excess penetrant

It is now necessary to remove excess penetrant from the surface of the component. Some penetrants can be washed off the surface with water, while others require the use of specific solvents. Uniform removal of excess penetrant is necessary for effective inspection.

Development

The development stage is necessary to reveal clearly the presence of any defect. The developer is usually a very fine chalk powder. This may be applied dry, but more commonly is applied by spraying the surface with chalk dust suspended in a volatile carrier fluid. A thin uniform layer of chalk is deposited on the surface of the component. Penetrant liquid present within defects will be slowly drawn by capillary action into the pores of the chalk. There will be some spread of penetrant within the developer and this will magnify the apparent width of a defect. When a dye penetrant is used the dye colour must be in sharp contrast to the uniform white of the chalk-covered surface. The development stage may sometimes be omitted when a fluorescent penetrant is used.

Observation and inspection

After an optimum developing time has been allowed, the component surface is inspected for indications of penetrant 'bleedback' into the developer. Dye-penetrant inspection is carried out in strong lighting conditions, while fluorescent-penetrant inspection is performed in a suitable screened area using ultra-violet light. The latter technique causes the penetrant to emit visible light, and defects are brilliantly outlined.

The five essential operations are shown in figure 2.1.

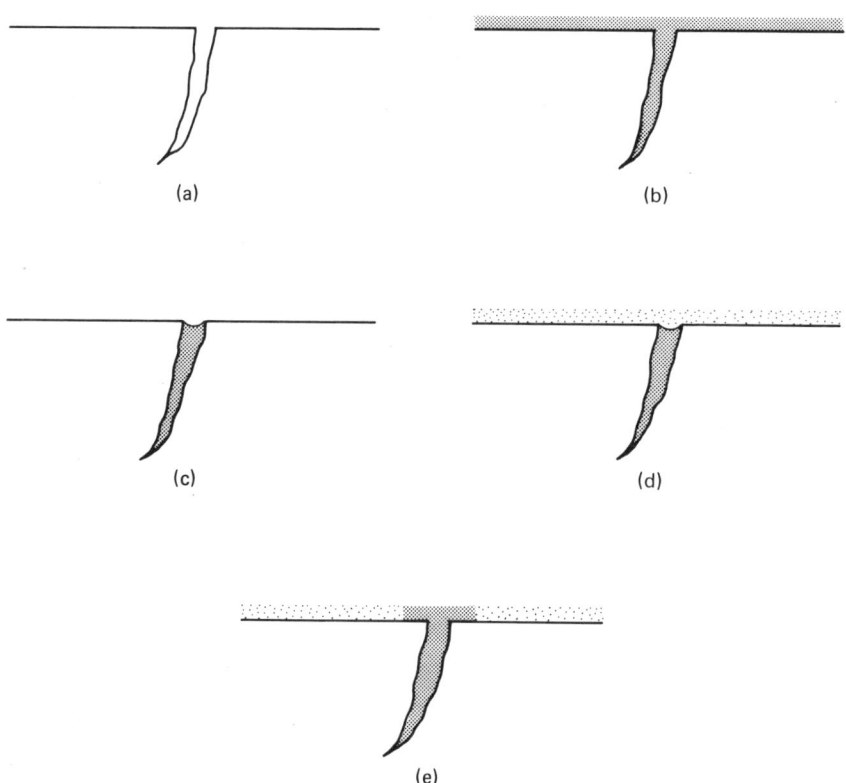

(a)

(b)

(c)

(d)

(e)

FIGURE 2.1   Stages in penetrant testing: (a) material surface clean and grease-free; (b) penetrant absorbed into defect; (c) excess penetrant removed, but liquid remains in defect; (d) developer applied to surface; (e) penetrant absorbed into developer giving indication of defect.

2.3 Characteristics of a penetrant

A liquid penetrant must possess certain specific characteristics if the inspection system is to be efficient. The various qualities required are listed below and penetrant formulations have been developed to give a balanced combination of the many parameters.

Penetration

The penetrant must have the ability to enter extremely fine surface defects or other openings in the component under test.

Body

The penetrant must have a good surface wetting ability and be able to maintain a surface film on the component, and hence, continue to feed into a defect over a considerable period of time.

Fluidity

In addition to the properties mentioned above, the penetrant should have the ability to drain away from the component well but with a minimum amount of dragging-out of penetrant from within defects.

Solution ability

If necessary, the penetrant should be capable of dissolving a path into contaminated defects, through a wide range of contaminants.

Stability

The liquid penetrant should be stable over a wide range of temperature and humidity and should not form a scum or lose its volatile constituents while it is kept in open tanks or when stored in drums.

Washability

It must be possible to remove excess penetrants from component surfaces easily without affecting the penetrant within any defects.

Drying characteristics

A penetrant must resist drying out, and complete bleed out, during hot-air drying of the component after the wash operation has been completed. Ideally, heat should aid in promoting a return of the penetrant to the component surface, in order to produce a sharply defined indication.

Visibility

The dye used in dye penetrants should be such that a good, deep colour can be given to the penetrant by a comparatively small amount of dye. If a large amount of dye is used in the formulation, the penetrating abilities of the fluid could be reduced. Red is the most commonly used colour in dye penetrants as this colour is the most readily seen by the human eye.

In addition to the characteristics listed above, it is important that any penetrant should be safe to use. If the operational method involves the dipping of components in a bath of penetrant then it is desirable that the fluid kept in large open tanks has a high flash point and certainly not less than 60°C. However, penetrants of lower flash points can be tolerated if the method of application is by spraying or brushing on to small surface areas. Preferably the penetrants used should be non-toxic but this is not always possible. In those cases where there may be health problems from vapour inhalation or the effects of liquid on skin, the safety precautions listed by the penetrant manufacturers must be adhered to.

A number of penetrant types, or classes, have been developed to cater for the wide variety of inspection conditions that can occur in practice. The main types of penetrant systems currently in use are water-washable systems, post-emulsification systems, and solvent systems.

2.4 Water-washable system

This system (using a fluorescent or visible dye penetrant) is designed so that the penetrant can be directly removed from the component surface by washing with water. The processing is thus rapid and efficient. It is extremely important, however, to maintain a controlled washing operation, especially where the removal of excess penetrant is by means of water sprays. A good system will be an optimisation of the processing conditions, such as, water pressure and temperature, duration of rinse cycle, surface condition of the workpiece, and the inherent removal characteristics of the penetrant. Even so, it is possible that penetrant may be washed away from small defects.

2.5 Post-emulsification system

When it is necessary to detect minute defects, high-sensitivity penetrants that are not water washable are usually employed. Such penetrants have an oil base, and require an additional processing step. An emulsifier is applied after the penetrant has had sufficient time to be absorbed into defects. The major advantage of this system is that the emulsifier renders the excess penetrant soluble in water, and hence, capable of being rinsed away. Provided the process is carefully controlled, any penetrant within flaws is not affected. Small flaws that are often missed because of the penetrant being washed out are more likely to be identified by the use of this system.

Despite this advantage, the system is more expensive, because of the penetrant-emulsifier costs and the additional time required for the operations. The use of this system also necessitates additional handling equipment and space.

2.6 Solvent-removable system

It is often necessary, to inspect only a small area of a component, or to inspect a component *in situ* rather than at a regular inspection station. Solvent-removable penetrants are widely used for such situations. Normally, the same type of solvent is used both for pre-cleaning and for the removal of excess penetrant.

There are two basic solvent types: flammable and non-flammable. The flammable cleaners are potential fire hazards but are free from halogens, while the non-flammable cleaners are halogenated solvents, which render them unsuitable for use in confined spaces, because of their high toxicity.

The excess surface penetrant is usually removed by wiping the component with a lint-free cloth moistened with solvent. Flooding techniques should be avoided, if possible, since there is a high risk of removing penetrant from the flaws. When carried out with due care and consideration, this system is extremely sensitive. However, the costs are relatively high, because of high material expense and the fact that it is a more labour-intensive process.

2.7 Surface preparation and cleaning

In order to obtain the most accurate results with liquid penetrants it is necessary that the surface of the workpiece be thoroughly cleaned. If the surface is not properly cleaned the presence of defects may be missed because the penetrant may not be able to enter the flaw or, if the area of surface immediately surrounding a defect is contaminated the true appearance of the defect may be masked because of retention of the penetrant by surface contaminants. There is also the possibility that the penetrant may interact with some forms of surface contamination and, in consequence, its ability to enter fine cracks may be reduced.

A variety of methods may be used, either separately or jointly, to clean the surface of components thoroughly prior to the application of penetrant, and the methods chosen will depend on the nature of the components, the type of surface contamination and the number of components which have to be inspected. Wire brushing, grit blasting, either wet or dry, or abrasive tumbling can be used to remove light oxide coatings, adherent welding flux and dirt deposits. Ultrasonic cleaning is a suitable process when large numbers of small components are involved. Oils and greases are best removed using solvents or high-pressure water and steam cleaning. Chemical cleaning techniques can also be used. Oils, greases and carbon deposits may be removed with alkaline solutions, while strong acid solutions may be employed to remove thick oxide scales.

2.8 Penetrant application

In practice, penetrants may be applied to the surface of the component by one of several methods. The method chosen will depend on the size, shape and the number of parts to be inspected. It will also depend on whether or not components are to be examined *in situ*.

When large numbers of comparatively small parts are to be examined, complete immersion of the components in a tank containing liquid penetrant is generally the preferred method. The components must be completely dry prior to immersion, since water or residual cleaning solvents will inhibit penetration and contaminate the penetrant. During component immersion, care must be exercised to ensure that air pockets are avoided, and that all surfaces to be examined are completely wetted. Usually, components are dipped for a pre-determined period and then drained. During this stage, care must be taken to ensure that the penetrant drains from all cavities. Components which exhibit residual penetrant on their surfaces, after drying should be re-dipped.

Flooding is normally employed to examine large areas of single components. Usually, the penetrant is applied with a low-pressure spray which does not result in fluid atomisation. Care must be taken to ensure that the penetrant covers the whole of the surface to be examined, and that the surface remains wet for the whole of the penetration period.

When it is only required to examine individual components or to inspect components *in situ*, the penetrant is applied by brush or from an aerosol spray can. Brush application is preferable if the component is of complex shape. As with flood application, the penetrant should not be allowed to dry on the surface.

The length of time that the penetrant is in contact with the component is important. Penetrant will seep into fairly large flaws in a few seconds but it may take up to 30 minutes for the liquid to penetrate into very small defects and tight cracks. The penetration time that is used in practice will be determined by the nature and size of the defects which are being sought. Depending on the application, penetration times of between 20 seconds and 30 minutes are used.

2.9 Development

The developer is a critical part of an inspection process. Borderline indications that might otherwise be missed can be made readily visible by using a suitable developer. Moreover, the employment of such materials reduces inspection time by hastening the appearance of indications.

In order to attain optimum inspection conditions, developers are designed to work with specific penetrants. Hence, a developer which has been 'tuned' to a specific penetrant should be confined to use with that particular penetrant, since it may be totally ineffective with other penetrant media.

In order to perform its function a developer should possess an optimum combination of the following characteristics:

Absorption A developer must be easily wetted by the penetrant at the flaw, and highly absorptive to draw the maximum amount of penetrant from the defect.

Application It must be easy to apply and capable of forming a thin uniform surface coating. In addition, it must be easy to remove after inspection.

Background masking It must be capable of effectively masking out interference from background colours, and capable of providing a contrasting background for indications, especially when coloured penetrants are used.

Physical characteristics It must have a grain size and a particle shape that will disperse the penetrant at the flaw, so that a clearly defined indication is attained, without excessive spread. In addition, the material must be neither hygroscopic nor excessively dusty.

Chemical characteristics It must not contain ingredients which may be harmful to either the parts being inspected or to the operator.

If it takes a long time for penetrant to be drawn into a tight crack then it follows that a similar length of time will be needed for liquid to be drawn from the defect by the developer. In consequence, development times of between 10 and 30 minutes are generally required to ensure that all defect indications are visible.

2.10 Advantages and limitations

The liquid penetrant process is comparatively simple as no electronic systems are involved, and the equipment necessary is cheaper than that required for other non-destructive testing systems. The establishment of procedures, and inspection standards for specific product parts, is usually less difficult than for more sophisticated methods. The technique can be employed for any material except porous materials, and, in certain cases, its sensitivity is greater than that of magnetic particle inspection. Penetrant inspection is suitable for components of virtually any size or shape and is used for both the quality control inspection of semi-finished and finished production items and for routine in-service inspection of components. This latter may be *in situ* inspection, thus obviating the need to dismantle a large complex assembly, or it may be the inspection of dissembled components during, say, the major overhaul of an aircraft.

The obvious major limitation of a liquid penetrant system is that it can detect surface-breaking defects only. Sub-surface defects require additional inspection methods. Other factors inhibiting the effectiveness of liquid penetrant inspection are surface roughness and porous materials. The latter, in particular, can produce false indications, since each pore will register as a potential defect.

2.11 Range of applications

The range of applications of liquid penetrant testing is extremely wide and varied. The system is used in the aerospace industries by both producers for the quality control of production and by users during regular maintenance and safety checks. Typical components which are checked by this system are turbine rotor discs and blades, aircraft wheels, castings, forged components and welded assemblies. Many automotive parts, particularly aluminium castings and forgings, including pistons and cylinder heads, are subjected to this form of quality control inspection before

FIGURE 2.2   Inspection of inlet assembly of aircraft gas turbine using ultra-violet light after application of fluorescent penetrant (courtesy of Magnaflux Ltd).

assembly. Penetrant testing is also used for the regular in-service examination of the bogie frames of railway locomotives and rolling stock in the search for fatigue cracking.

Figure 2.2 shows the inspection of the inlet assembly of an aircraft engine. The operator is looking for surface cracks under ultra-violet light, after the assembly has been processed with fluorescent penetrant. In general manufacturing industry, the techniques are widely used for the process and quality control checking of castings, forgings and weldments. Figure 2.3 shows a typical industrial processing line for inspection using a water-washable fluorescent penetrant system. The installation comprises a penetrant tank, water wash, drier, development station and a hooded inspection booth for viewing under ultra-violet light.

A convenient method of applying penetrant to large components is by spraying, and when it is required to inspect large numbers of large components it is convenient to use a spraying booth. Figure 2.4 ahows an electrostatic spray booth in use. The casting is given an electrostatic charge and the penetrating liquid is charged, but with the opposite sign, as it leaves the spray gun. This ensures that all

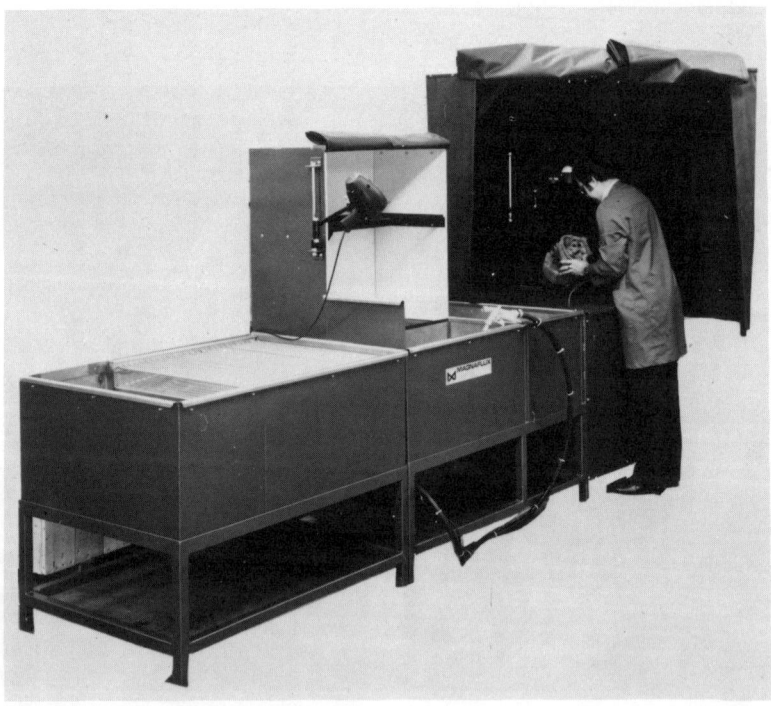

FIGURE 2.3   Fluorescent penetrant processing line incorporating penetrant tank, wash water, drier, development and ultra-violet light inspection stations (courtesy of Magnaflux Ltd).

parts of the testpiece receive a uniform coating of penetrant with the minimum wastage of penetrant. Other applications of penetrant testing include the detection of glazing cracks in electrical ceramic parts such as spark plug insulators, the search for cracks in glass-to-metal seals in electrical components and the detection of flaws in moulded plastic parts.

FIGURE 2.4   Electrostatic fluorescent penetrant spray booth (courtesy of Magnaflux Ltd).

3

Magnetic Particle Inspection

3.1 Introduction

Magnetic particle inspection is a sensitive method of locating surface and some sub-surface defects in ferro-magnetic components. The basic processing parameters depend on relatively simple concepts. In essence, when a ferro-magnetic component is magnetised, magnetic discontinuities that lie in a direction approximately perpendicular to the field direction will result in the formation of a strong 'leakage field'. This leakage field is present at and above the surface of the magnetised component, and its presence can be visibly detected by the utilisation of finely divided magnetic particles. The application of dry particles or wet particles in a liquid carrier, over the surface of the component, results in a collection of magnetic particles at a discontinuity. The 'magnetic bridge' so formed indicates the location, size, and shape of the discontinuity.

Magnetisation may be induced in a component by using permanent magnets, electro-magnets or by passing high currents through or around the component. The latter technique is widely used for production quality control applications because high-intensity magnetic fields can be generated within components. Hence, good sensitivity in flaw indication and detection is attained.

3.2 Magnetisation

The direction of a magnetic field in an electro-magnetic circuit is controlled by the direction of current flow. The magnetic lines of force are always at right angles to the direction of the current flowing in a conductor. The simple right-hand corkscrew rule describes the field/current orientation relationship.

Current passing through any straight conductor such as a wire or bar creates a circular magnetic field around the conductor. When the conductor is a ferro-magnetic material, the current induces a magnetic field within the conductor as well as within the surrounding space. Hence, a component magnetised in this manner is circularly magnetised, as shown in figute 3.1a.

An electric current can also be used to create a longitudinal magnetic field in components. When current is passed through a coil of one or more turns surrounding a component, a longitudinal magnetic field is generated within the workpiece as shown in figure 3.1b.

The effectiveness of defect indication will depend on the orientation of the flaw to the induced magnetic field and will be greatest when the defect is perpendicular to the field.

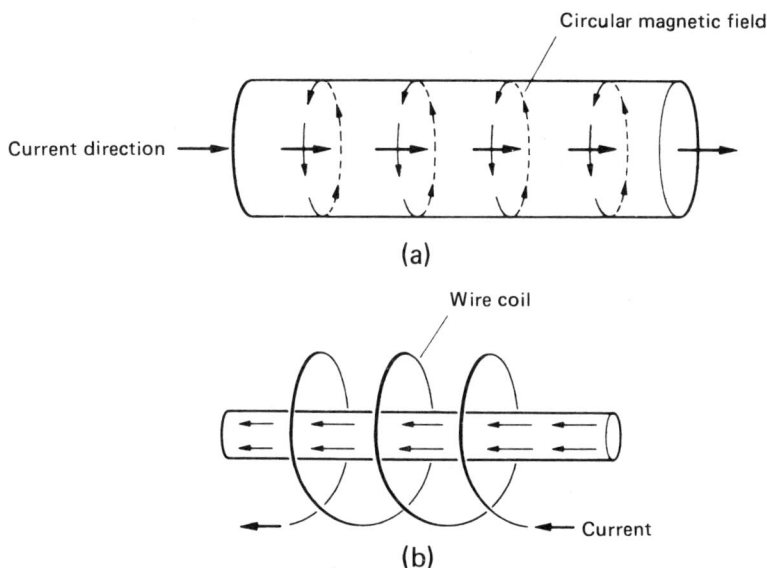

FIGURE 3.1   (a) Current passed through workpiece, inducing circular magnetisation. (b) Longitudinal magnetisation induced by placing workpiece within a coil.

This is shown schematically in figure 3.2. In this, defect A will give a strong indication. Defects B and C are sub-surface defects. An indication will be obtained in respect of defect B as it is normal to the magnetic field but defect C will not be indicated. In the case of defect C it is too deep within the material to give an indication but, even if such a defect were close to the surface its alignment with the magnetic field would render detection unlikely.

Generally, in order to indicate the presence of all flaws a component will need to be magnetised more than once. For components of relatively simple shape this is achieved by, firstly, inducing circular magnetisation to indicate longitudinal defects. The component is then demagnetised before magnetising for a second

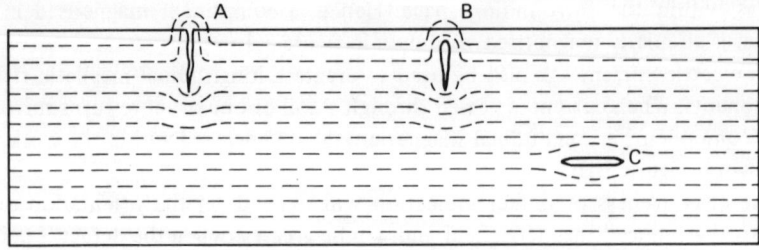

FIGURE 3.2   Magnetic flaw detection. Detectable surface leakage fields pro-
              duced by defects A and B. Defect C likely to remain undetected.

time within a coil to induce longitudinal magnetisation which will enable transverse
defects to be located. The necessity for a two-stage magnetisation procedure is
obviated if the 'swinging field' technique for magnetisation is used. This involves
the application of a three-phase magnetising current. The result of this is to
establish a swinging or rotating magnetic field within the component which
permits defects in all orientations to be detected with one magnetisation.

Figure 3.3 gives a representation of a bar containing a number of defects. All
the defects will be indicated if the bar is given both circular and longitudinal
magnetisation.

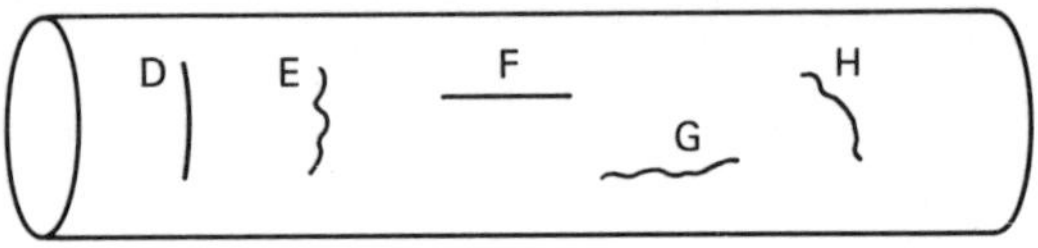

FIGURE 3.3   Bar containing several surface defects.

Defect D will give no leakage field indication with circular magnetisation but
the more irregular defect E may give a weak indication if the field strength is
sufficiently high. However, when the component is magnetised longitudinally
both D and E will be clearly indicated. Similarly, defect F will not be detectable
using longitudinal magnetisation but the less regular defect G may give a weak
indication while both F and G will show up clearly to circular magnetisation. A
defect at some angle, such as defect H, should be seen on both occasions.

When a component is of complex shape, the induced magnetic fields will be
distorted and will often result in a combination of both circular and longitudinal
magnetisation.

3.3 Magnetisation methods

Magnetisation of a component could be accomplished using permanent magnets but generally magnetic fields are induced by passing a heavy current through the component, by placing a coil around or close to the component under test, or by making the component part of a magnetic circuit, for example by means of a hand yoke as shown in figures 3.4e and 3.8.

The actual method used will depend on the size, shape and complexity of the parts to be inspected and in components to be inspected *in situ*, the accessibility to such components.

Direct electrical contact at each end of a component, so that current passes through the whole part (see figure 3.4a) is a rapid and reliable method which is very suitable for the inspection of relatively small components. Circular magnetisation is produced over the whole length of the workpiece and a good sensitivity can be achieved. This type of inspection is generally made using purpose-built test equipment with the workpiece held in the horizontal mode between adjustable contact clamps and in conjunction with 'wet' magnetic particles.

Small to medium size components in which one dimension, namely length, predominates can be readily magnetised in a longitudinal direction by placing within a coil (figure 3.4b). The part must be held centrally within the coil to achieve the best results. This technique is particularly useful for the location of transverse flaws in such items as axles, crankshafts and camshafts. Large castings or forgings can be magnetised in the longitudinal direction by winding a flexible cable around them.

It may be necessary to repeat the process more than once, placing the winding in different positions, to obtain complete inspection of a large and complex part.

Flexible cables with prod contact pieces are widely used for the inspection of large castings and forgings. If the contacts are placed at opposite ends of a large component, the entire piece will be magnetised and inspection will be completed in a short time. However, in the case of a large workpiece a very large current running into thousands of amperes may be required if the degree of magnetisation necessary to indicate defects clearly is to be achieved. This would require the use of expensive high-duty electrical equipment. Alternatively, the prod contacts can be placed relatively close to one another to examine a small area of the component, and then the operation repeated until the complete part has been covered (figure 3.4c). The time taken for this form of inspection will, of course, be longer than if the entire component is magnetised at once but the electrical power requirement is very much less. It may be that there is only a requirement to inspect a small portion of a structure, for example, a weld. In this case the prod contact method is very useful. The two contacts can be placed close to the weld and by this method it is possible to detect cracks, lack of weld penetration and, in some cases, inclusions. When using prod contacts, care must be taken to ensure good electrical contact otherwise arcing may occur between the prod and the workpiece surface, resulting in over-heating and burn damage.

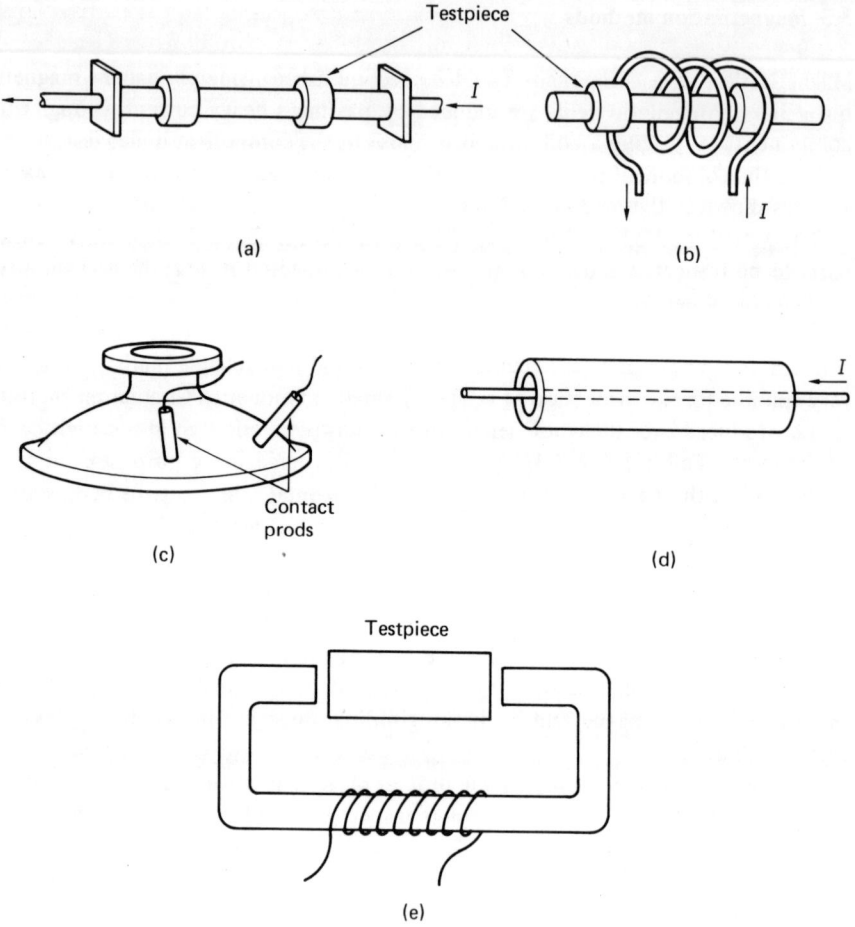

FIGURE 3.4   Magnetising methods: (a) current passed through complete part,
              inducing circular magnetisation; (b) part placed within coil,
              inducing longitudinal magnetisation; (c) prod contacts placed on
              surface of large castings; (d) hollow section magnetised by thread-
              ing a conducting cable through it; (e) part magnetised within
              magnetic yoke.

A component with a continuous hole through it can be magnetised by energis-
ing a straight conducting cable passing through the hole (see figure 3.4d). This
inspection technique is often used in the examination of parts such as pipe con-
nectors, hollow cylinders, gear wheels and large nuts.

For certain types of application an electro-magnetic yoke may be used (figure
3.4e).

This technique is suitable for the examination of a variety of shapes. One example of this method is the search for forging laps and other surface defects in crane hooks. Good sensitivity can be achieved but it is important that the yoke be positioned correctly in relation to the orientation of the anticipated flaws.

The current supply used for all the above mentioned methods may be either AC or DC. When an AC supply is used the magnetic flux generated within the workpiece is largely confined to the area close to the material surface but flux penetration into the material is deeper with DC excitation. Hence, if a major aim of the inspection is to detect sub-surface as well as surface discontinuities the use of DC is to be preferred. Good results are generally obtained with half-wave rectification. One advantage of using AC is that it is easy to demagnetise the component after test by gradually reducing the current level to zero.

3.4 Continuous and residual methods

Irons and steels have differing magnetic characteristics. Pure iron and low carbon steels in the annealed or normalised conditions are magnetically 'soft' and are of low coercivity and possess low remanence. In other words they can be magnetised easily but also lose most of their magnetism quickly when the magnetising field is removed. Conversely, many alloy steels and hardened steels are magnetically 'hard'. They are more difficult to magnetise but possess high remanence, retaining much of the induced magnetism when the magnetising field is removed.

Soft materials of low remanence must be tested using the *continuous method*. In this the dry or wet magnetic particles used to indicate the presence of discontinuities are applied to the component while the magnetising current is flowing through the component or magnetising coil. The current may be continuous or it may be applied as a series of short 'shots'. The continuous method is very sensitive and will give indications of very fine defects. However, it should be emphasised that when used in conjunction with wet particles, particle application should cease before the magnetising current is switched off, otherwise the particles could be washed away and no indications observed.

When a material possesses a high remanence the component may be magnetised, the field removed and then the magnetic particles applied and inspection carried out. This is known as the *residual method*. It has the advantage that inspection may be made away from the magnetising equipment, if necessary. The sensitivities possible in the residual method are generally less than those of the continuous method.

3.5 Sensitivities

Magnetic particle inspection can be a highly sensitive technique but there are several factors which will affect the sensitivity. A major factor, which has already been mentioned, is the orientation of the discontinuity with respect to the induced magnetic field, and sensitivity will be greater when the flaw lies at right angles to the field. Other major factors are the size, shape and general characteristics of the magnetic particles used and also the nature of the carrying fluid for these particles (this is discussed in section 3.7).

The strength of the magnetic field is also a factor affecting sensitivity and generally the sensitivity increases as the field strength increases, but there is a limit to this effect as with very high field strengths magnetic particles will be attracted to crack-free areas of the surface as well as to flaws. The actual shape of components also affects sensitivity and, in practice, the optimum field strength for a component of a particular shape is determined by trial and error.

In the best conditions it is possible to detect cracks with a width of as little as 10^{-3} mm. The depth below the surface at which sub-surface defects may be detected is of the order of 3 to 7 mm when magnetisation by DC current is used, but this will also depend on the size, shape and orientation of the flaw. With AC magnetisation it is doubtful if sub-surface defects would be detected unless they were within one millimetre of the surface.

3.6 Demagnetisation

The amount of magnetism which remains in a ferro-magnetic material after magnetic particle inspection will depend upon the physical characteristics of the material. As stated earlier, magnetically 'hard' materials will have a high remanence. It will often be necessary to demagnetise a component after magnetic particle inspection. There are several reasons for this. The component may be intended for use in an area where a residual magnetic field will interfere with the operation or the accuracy of instruments that are sensitive to magnetic fields. Another reason is that abrasive particles may be attracted to magnetised parts, such as bearing surfaces, bearing raceways and gear wheels, causing accelerated wear damage. Particles could adhere to a magnetised surface and interfere with subsequent operations, such as painting or electroplating. If a magnetised part is machined, chips could adhere to the surface being machined and adversely affect the surface finish, dimensions and tool life. Finally, during any subsequent electric arc welding operations, strong residual magnetic fields could deflect the arc from its point of application.

The principle of demagnetisation is to subject the magnetised component to a field which continually reverses direction and at the same time gradually decreases in strength to zero. Figure 3.5 shows hypothetical current/flux density curves during demagnetisation.

When AC equipment has been used to magnetise components for inspection, the same equipment can be used for demagnetising. When inspection is complete the magnetising current is gradually reduced to zero with the workpiece still clamped between the electrical contacts, or while it is still positioned within a

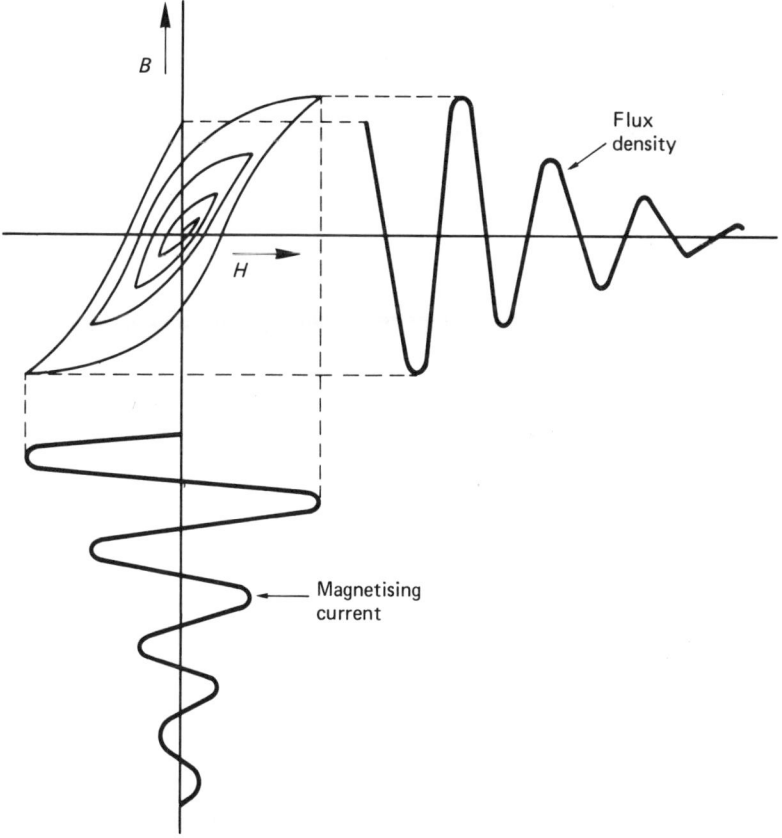

FIGURE 3.5   Current and flux density curves during demagnetisation.

magnetising coil. If the initial magnetisation was induced by a DC current, additional AC equipment is required for demagnetisation purposes. Components of large cross-sectional size are difficult to demagnetise fully using a diminishing AC current, as AC effects tend to be confined to surface layers. In these cases it is usual to combine AC and DC demagnetisation.

Demagnetisation can also be carried out using an AC yoke. The space between the poles of the yoke should be such that the part to be demagnetised will pass through them in close proximity. The poles of the yoke are placed on the component surface and a stroking action is employed. The strokes must always be in the same direction along the component, and the yoke should be moved away in a circle on the return stroke.

3.7 Magnetic particles

The magnetic particles which are used for inspection may be made from any ferromagnetic material of low remanence and they are usually finely divided powders of either metal oxides or metals. The particles are classified as *dry* or *wet* according to the manner in which they are carried to a component. Dry particles are carried in air or gas suspension while wet particles are carried in liquid suspension.

The normal carrier for dry particles is air and the particle cloud is produced using a mechanical powder blower, or a rubber spray bulb. Care must be taken and the powder should not be blown under pressure directly at the component surface because under these conditions the powder particles would not be free to be attracted by all leakage fields. The particles should reach the magnetised component in a uniform cloud. If dry particle inspection is to be carried out, it is important that the component surface be free from grease and other adhering deposits as the powder could be held by such contaminants and give a false defect indication. For good visibility, dry particles are available in several colours — the usual ones being yellow, red or black. Dry particles are also available with a fluorescent coating. (The components have to be visually inspected under ultra-violet light when this last mentioned powder is used.) Magnetite, Fe_3O_4, is a black powder. The addition of pigments to produce coloured powders tends to give a reduced sensitivity although, of course, the visibility of a coloured powder may be greater than that of a black powder. This will depend on the nature of the component surface.

Dry powder inspection is very useful in connection with portable magnetising equipment and is capable of giving good defect indications, particularly if the component surface is slightly rough. It also tends to be more sensitive than wet particle inspection for the indication of sub-surface defects.

Wet particles are normally employed in stationary equipment, where a bath can be used without any inherent handling problems. The liquid carrier is usually a light petroleum distillate such as kerosene, but it may be water. The particles are normally available in black and red pigments, or as blue-green or yellow-green fluorescent powder.

Wet particles possess better particle mobility and are easier to apply than dry particles. The presence of surface grease is less of a problem than with dry powder because the carrying fluid is generally petroleum based. However, special precautions against the hazards of toxicity and fire are necessary. In addition, the liquid bath must be constantly agitated to maintain freedom of particle movement, and regularly cleaned to remove contamination.

Very high sensitivities are possible with wet particle inspection, particularly when a fluorescent powder is used and inspection is made under ultra-violet light. Although actual particle size and shape depend mainly on a particular manufacturer's policy, it is necessary to be aware of the advantages and limitations imposed by these parameters.

Particle size

Coarse particles are better than fine particles for bridging large gaps or cracks, but fine particles will give a better sensitivity to small defects. This is because coarse particles are not likely to be arrested by weak leakage fields, while fine particles are easily arrested by very weak fields. Fine particles tend to adhere to fingerprints, soiled areas and rough surfaces, and this could lead to false indications and the possible obscuring of true defects.

Long, slender particles develop a stronger polarity than spheroidal particles and can therefore produce indications of defects more readily but they suffer from the disadvantage that they may bind together in uneven clumps, reducing particle mobility. Strong indications of defects with a high sensitivity can be obtained when the magnetic particles are composed of a blend with some of spheroidal form and some possessing an elongated shape.

3.8 Applications

The principal industrial uses of magnetic particle inspection are in-process inspection, final inspection, receiving inspection and in maintenance and overhaul.

Although in-process inspection is used to highlight defects, as soon as possible in the processing route, a final inspection gives the customer a better guarantee of defect-free components.

During receiving inspection, both semi-finished purchased parts and raw materials are inspected to detect initial defects. Incoming rod and bar stock, forging blanks and rough castings are inspected in this way.

The transportation industries (road, rail, aircraft and shipping) maintain planned overhaul schedules at which critical parts are inspected for cracks. Crankshafts, frames, flywheels, crane hooks, shafts, steam turbine blades and fasteners are examples of components vulnerable to failure, in particular fatigue failure. Hence, there is a need for regular inspection.

Figure 3.6 shows an internal combustion engine crankshaft in position in a magnetic test unit. The shaft is clamped between electrical contacts, which when energised will cause circular magnetisation. The illustration also shows the magnetising coil which is used to induce longitudinal magnetisation. Figure 3.7 shows an installation for the magnetic particle inspection of pipe couplings. The horizontal rod visible in the illustration is a conductor. The section of pipe will be positioned symmetrically about the conductor before a magnetising current is passed through the rod. A small adjustable magnetic yoke is shown in Figure 3.8. This is suitable for use with a wide variety of components. The adjustable articulated legs of the yoke will open up to a maximum separation of 250 mm.

Where there is a requirement for 100 per cent inspection of large production quantities of small items, the magnetic particle inspection system can be readily

FIGURE 3.6   Test bench for magnetic particle inspection of crankshafts (courtesy of Magnaflux Ltd).

FIGURE 3.7   Central conductor unit for the magnetic particle inspection of pipes and pipe couplings (courtesy of Magnaflux Ltd).

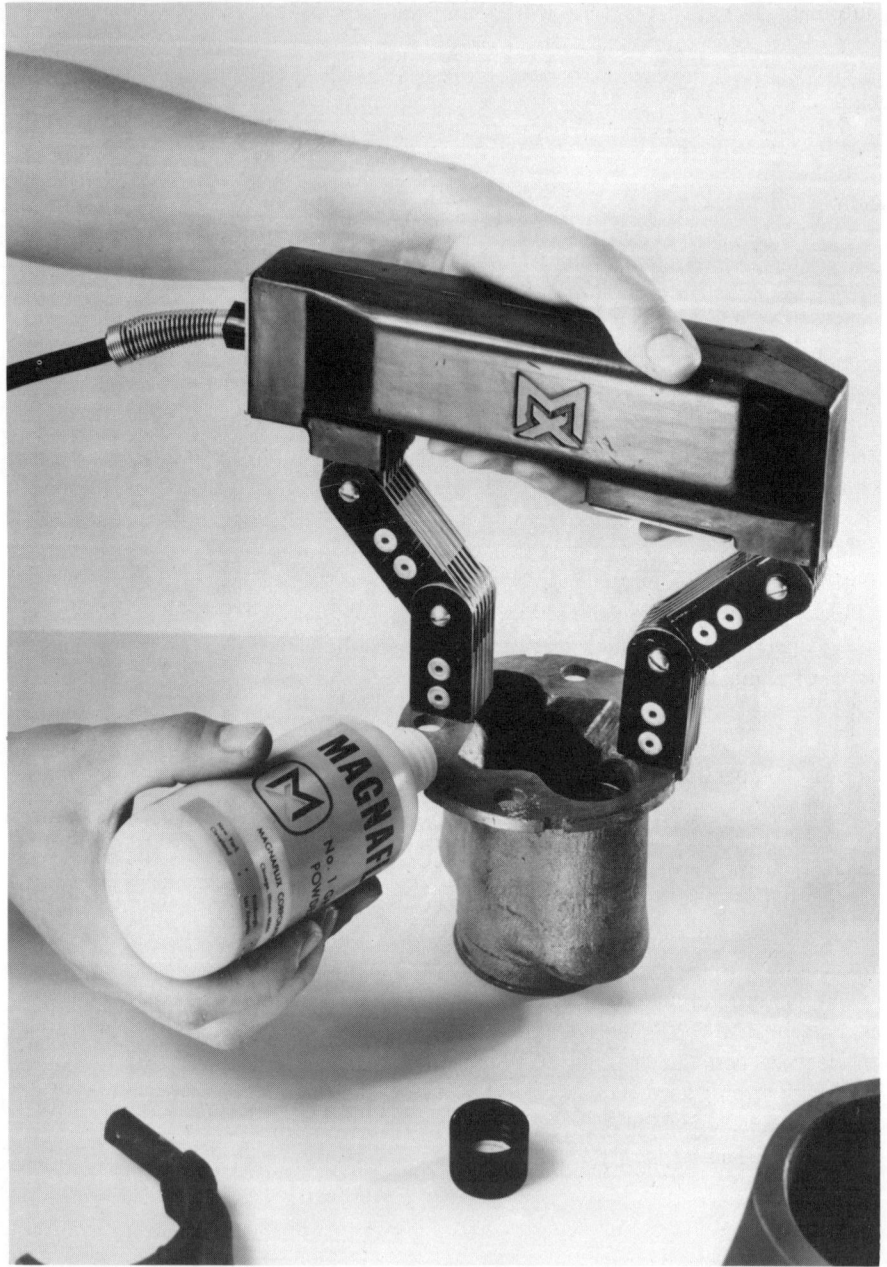

FIGURE 3.8   Small magnetic yoke with adjustable articulated legs. The operator is applying a colour contrast dry magnetic powder (courtesy of Magnaflux Ltd).

automated. Loading, conveying, magnetisation, manipulation and demagnetisation can all be fully automated, which results in consistent and relatively effortless inspection. The inspector can be devoted full time to inspection while the processing system performs more mundane functions. Production rates are met by having the various functions performed simultaneously at different stations. The cost of automation can be justified when compared with the cost of manning a large number of single manual units, necessary to achieve comparable production rates.

Automated inspection is used for ball and roller bearings, bearing races and rings, small castings and forgings, couplings, crankshafts and steel-mill billets.

3.9 Advantages and limitations of magnetic particle inspection

Magnetic particle inspection is a sensitive means of detecting very fine surface flaws and in certain situations it is superior to more sophisticated techniques in this respect. It is also possible to obtain indications from some discontinuities that do not break through the surface provided that they are close to the surface. It is often unnecessary to have an elaborate pre-cleaning routine and it is sometimes possible to obtain good indications even if a flaw contains contaminating material. While magnetic particle inspection is not a quantitative test, a skilled and experienced operator may be able to give an estimate of the breadth and depth of cracks. It is, of course, possible to determine accurately the depth of a crack which has been discovered in a magnetic test using some other technique (refer to chapter 7). A further advantage of magnetic particle inspection is that the equipment necessary is comparatively cheap and there is little needed in the way of ancillary equipment.

The major limitations of the technique are that it is only suitable for ferromagnetic materials and that, for the best results, the induced magnetic field should be normal to any defect; thus, two or more magnetising sequences will be necessary, except when the 'swinging field' technique is used, and a demagnetising procedure will need to be carried out for many components after inspection. When large components are to be inspected, extremely large currents are required and care will be needed to avoid localised heating and surface burning at the points of electrical contact. The indications observed in magnetic particle inspection may be readily visible but, frequently, considerable reliance must be placed on the skill and experience of the operator for the correct interpretation of the significance of indications. The sensitivity of magnetic particle inspection is generally very good but this will be reduced if the surface of the component is covered by a film of paint or other non-magnetic layer.

4

Electrical Test Methods (Eddy Current Testing)

4.1 Introduction

The basic principle underlying the electrical test methods is that electrical eddy currents and/or magnetic effects are induced within the material or component under test and, from an assessment of the effects, deductions can be made about the nature and condition of the testpiece. These techniques are highly versatile and, with the appropriate equipment and test method, can be used to detect surface and sub-surface defects within components, determine the thickness of surface coatings, provide information about structural features, such as crystal grain size and heat treatment condition, and also to measure physical properties including electrical conductivity, magnetic permeability and physical hardness.

In the case of ferro-magnetic materials there is a continuous spectrum of effects ranging from predominantly magnetic effects at low frequencies to eddy current effects at the higher frequencies, where the magnetic effect is suppressed. At the higher-frequency end, the techniques used relate to assessment of the distortion and reduction of the eddy current fields induced within the material. Changes in eddy current fields will indicate those defects which affect the flow of eddy currents in the surface layers of the material, such as cracks. At low frequencies it is the magnetic effects which predominate and the effect that the material has on the B-H loop is observed, and this relates to structural properties such as hardness. With non-magnetic materials, only eddy current effects occur, irrespective of the frequency but, in general, inspection techniques for non-magnetic materials use the higher frequencies, that is greater than 1 kHz.

As it is not necessary for there to be direct electrical contact with the part under test, so the methods can be adapted for many applications, including the high-speed automatic inspection and sorting of materials.

From the above it might at first appear that a single piece of eddy current test equipment could be the answer to every quality control inspectors' prayer, but things are not quite as simple as that. Eddy current techniques are based on an indirect measurement system and it is necessary to establish clearly the relationship between the structural and geometrical characteristics of a testpiece and the instrument responses. It may be that, under certain circumstances, an instrument response caused by a change in testpiece dimensions, such as the presence of a change in cross-section or a keyway, could be confused with the type of response produced by a defect. However, with the adoption of the correct method for the particular application, an operator will be able successfully to detect defects or determine the structural characteristics which are being sought.

4.2 Principles of eddy current inspection

If a coil carrying an alternating current is placed in proximity to a conductive material, secondary or eddy currents will be induced within the material. The induced currents will produce a magnetic field which will be in opposition to the primary magnetic field surrounding the coil. This interaction between fields causes a back e.m.f. in the coil and, hence, a change in the coil impedance value. If a material is uniform in composition and dimensions, the impedance value of a search coil placed close to the surface should be the same at all points on the surface, apart from some variation observed close to the edges of the sample. If the material contains a discontinuity, the distribution of eddy currents — and their magnitude — will be altered in its vicinity and there will be a consequent reduction in the magnetic field associated with the eddy currents, so the coil impedance value will be altered.

Eddy currents flow in closed loops within a material and both the magnitude and the timing or phase of the currents will depend on a number of factors. These factors include the magnitude of the magnetic field surrounding the primary coil, the electrical and magnetic properties of the material, and the presence or otherwise of discontinuities or dimensional changes within the material. Several types of search coil are used, two common types being the flat or pancake type coil which is suitable for the examination of flat surfaces, and the solenoid type coil which can be used in conjunction with solid or tubular cylindrical parts. For tubes, a solenoid type coil may be placed around the tube or inserted into the bore. The patterns of eddy currents obtained with these coil types is shown in figure 4.1.

If a component contains a crack or other discontinuity, the flow pattern of eddy currents will be altered and this will cause change in the magnetic field and, hence, a change in coil impedance. A schematic representation of the effect of a discontinuity on eddy current pattern is shown in figure 4.2.

The impedance of a coil can be determined by measuring the voltage across it. In eddy current test equipment, changes in coil impedance can be indicated on a meter or a chart recorder or displayed on the screen of a cathode ray tube.

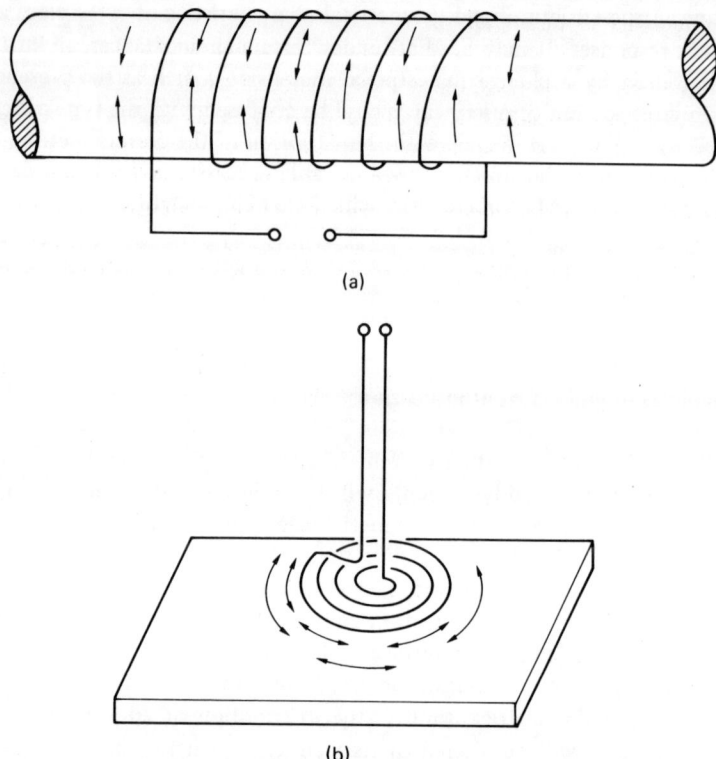

FIGURE 4.1   (a) Solenoid-type coil around bar, producing circumferential
              eddy currents. (b) Pancake-type coil, producing circular eddy
              currents within flat plate.

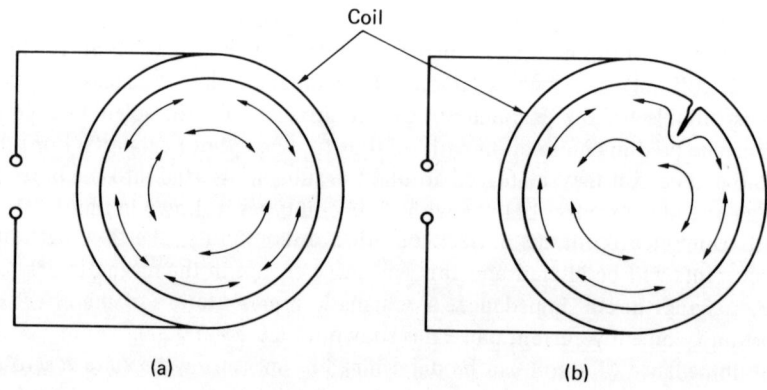

FIGURE 4.2   Cross-section of a bar within solenoid-type coil, showing eddy
              current pattern: (a) defect-free section — uniform eddy currents;
              (b) eddy current pattern distorted by the presence of a defect.

4.3 Conductivity of a material

The conductivity of a material is one of the major variables in eddy current testing. Each material has its own individual conductivity value. The electrical conductivity of a metal or alloy is affected by many factors. In many cases quite minor variations in chemical composition will alter the conductivity value. One effect of cold working on a metal is to decrease the electrical conductivity, and heat treatments will bring about changes. Any factor which will increase the physical hardness of a metal will cause a reduction in its electrical conductivity. See table 4.1.

Table 4.1 Variation of electrical conductivity with composition and condition for some copper materials and other metals

Composition	Amount of cold reduction (per cent)	Temperature (°C)	Conductivity (per cent of IACS value)
Copper with 0.04% P	0	20	72
Copper with 0.1% Fe	0	20	77
Copper with 0.45% As	0	20	41
	0	20	102.3
	10	20	100.8
	25	20	100.3
Copper with 0.002% Ag	50	20	99.7
	0	130	70.6
	10	130	69.8
	25	130	69.2
Silver	0	20	106
Aluminium	0	20	62
Iron	0	20	17

Conductivity is, of course, the inverse of resistivity. The resistivity of a material is a value with the units ohm-metre (Ωm) or micro-ohm-metre ($\mu\Omega$m) and the resistance of a piece of the material is related to its resistivity according to the relationship:

$$\text{Resistance } (\Omega) = \frac{\text{Resistivity } (\Omega\text{m}) \times \text{length (m)}}{\text{cross-sectional area } (\text{m}^2)}$$

Conductivity, therefore, has the units ohm-metre^{-1} $((\Omega\text{m})^{-1})$ or micro-ohm-metre^{-1} $((\mu\Omega\text{m})^{-1})$. The conductivity of a metal is generally expressed in one of two ways, either in micro-ohm-metre^{-1} or as a percentage of the conductivity of pure copper. The copper conductivity standard is a specific grade of high-purity copper known as the International Annealed Copper Standard (IACS). The resistivity of IACS copper is 1.7241×10^{-2} $\mu\Omega$m at 20°C. Alternatively it can be stated that the conductivity of IACS copper is 58.0 $(\mu\Omega\text{m})^{-1}$ and this value is taken as 100 per cent on the IACS scale.

4.4 Magnetic properties

A magnetic field, H, will produce a magnetic flux density, B. In a vacuum the magnetic flux density which is produced is related to the magnetising field strength as follows:

$$\frac{B}{H} = \mu_0$$

μ_0 is a constant and is termed the *magnetic permeability* of a vacuum. The value of μ_0 is $4\pi \times 10^{-7}$ henry per metre. The magnetic flux density induced in a material for any given magnetising field strength may differ from that induced in a vacuum and the relationship can be written

$$\frac{B}{H} = \mu_0 \mu_r$$

where μ_r is termed the *relative magnetic permeability*. For a vacuum, therefore, $\mu_r = 1.0$. The relative magnetic permeabilities of most materials differ only very slightly from 1.0 and for all practical purposes can be taken as unity.* However, ferro-magnetic materials possess very high values of relative magnetic permeability and the values of μ_r may range from 100 to 1 000 000.

The value of relative permeability is not constant for any particular ferro-magnetic material but varies with the magnitude of the magnetising force. The strength of any induced eddy currents will also vary considerably and will increase with an increase in relative permeability. From this it follows that the techniques used for the inspection of non-magnetic and ferro-magnetic materials may differ.

In a similar manner to electrical properties, the magnetic permeability of a material is also affected by variations in composition, hardness and microstructural condition. Small variations in permeability will have a greater effect on eddy current than small changes in electrical conductivity. This is particularly true at low test frequencies.

When a ferro-magnetic material is magnetised to saturation value, the magnetic permeability of the material will have a virtually constant value. The eddy current inspection of steels is often carried out with the part magnetised to saturation. In this way the influence of small variations of permeability due to microstructural differences can be almost eliminated and any variations in eddy current response will be an indication of flaws or other discontinuities. For this type of inspection the part to be inspected is placed within a coil carrying a direct current and, by adjusting the value of this direct current, magnetic saturation of the part can be achieved. The search coil is located within the magnetising coil and positioned close to the surface which is to be inspected.

---

*   The relative magnetic permeabilities of non-magnetic materials range from 0.99995 to 1.01.

4.5 Coil impedance

When an alternating current flows through a coil there will be two factors oppos-
ing the flow of current, namely the ohmic resistance, R, of the coil and the
inductive reactance, X_L. The inductive reactance X_L is given by

$$X_L = \omega L$$

where L is the inductance of the coil and $\omega = 2\pi f$, f being the frequency of the
alternating current.

The total resistance of the coil is termed the impedance, Z, and is given by

$$Z = \sqrt{(R^2 + \omega^2 L^2)}$$

and can be represented by the hypotenuse of a right-angled triangle (see figure 4.3).

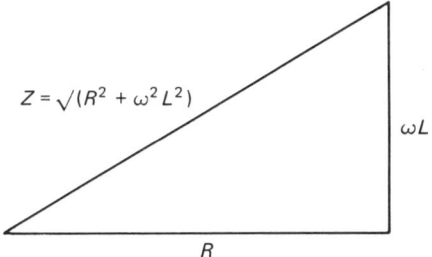

FIGURE 4.3   Impedance triangle.

The changes in impedance value of a coil can be represented on an impedance
plane diagram. The impedance of a coil can be represented on a graph with
reactance, ωL, and resistance, R, as the two axes (see figure 4.4).

A coil in air, with a reactance of ωL_0 and resistance R_0, can be represented by
point Z_0 on the diagram. When the coil is brought into close proximity to a conducting material the values L_0 and R_0 will be changed to ωL_1 and R_1, and so the
impedance value will change from Z_0 to Z_1. The magnitude and direction of this
change will be dependent on a number of factors. These factors are those related
to the nature of the testpiece, namely its dimensions, conductivity, magnetic
permeability and the presence of any defects, and those factors associated with
the test coil, namely the size and shape of the coil, its distance from the testpiece
and the frequency of the alternating source.

There is a lower limit to the frequency which can be used for any particular
inspection application and this is referred to as the *limit frequency*, f_g. The limit
frequency is not of constant value for all conditions and is dependent on the

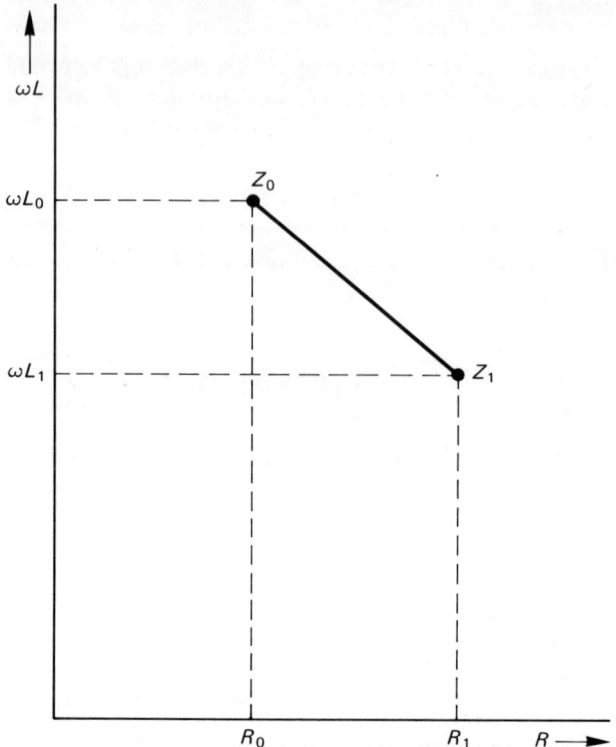

FIGURE 4.4   Impedance plane diagram. The impedance of the coil changes
from Z_0 to Z_1 when the coil is placed close to a conducting
material.

material parameters of conductivity and magnetic permeability and on the size
of the component. Forster and colleagues developed a mathematical analysis of
the behaviour of impedance in specimens of known geometrical shape. In the case
of a cylindrical testpiece within a solenoid type coil, the following empirical
relationship for limit frequency was derived:

$$f_g = \frac{506\,800}{\rho \mu_r d^2} \tag{4.1}$$

where p is the conductivity of the material $(\Omega m)^{-1}$, μ_r is the relative magnetic
permeability and d is the diameter of the testpiece in mm.

Coil impedance data may be plotted on impedance plane diagrams in order to
show the influence of the various parameters. It is customary to present this data
in the form of a normalised impedance plane diagram in which the axes are
$\omega L / \omega L_0$ and $R / \omega L_0$ rather than ωL and R.

For all frequencies other than the limit frequency, f_g, the ratio f/f_g can be calculated and figure 4.5 shows the variation of the resistance and reactance of a coil with changes in the frequency ratio when the coil is in conjunction with a specimen of constant conductivity.

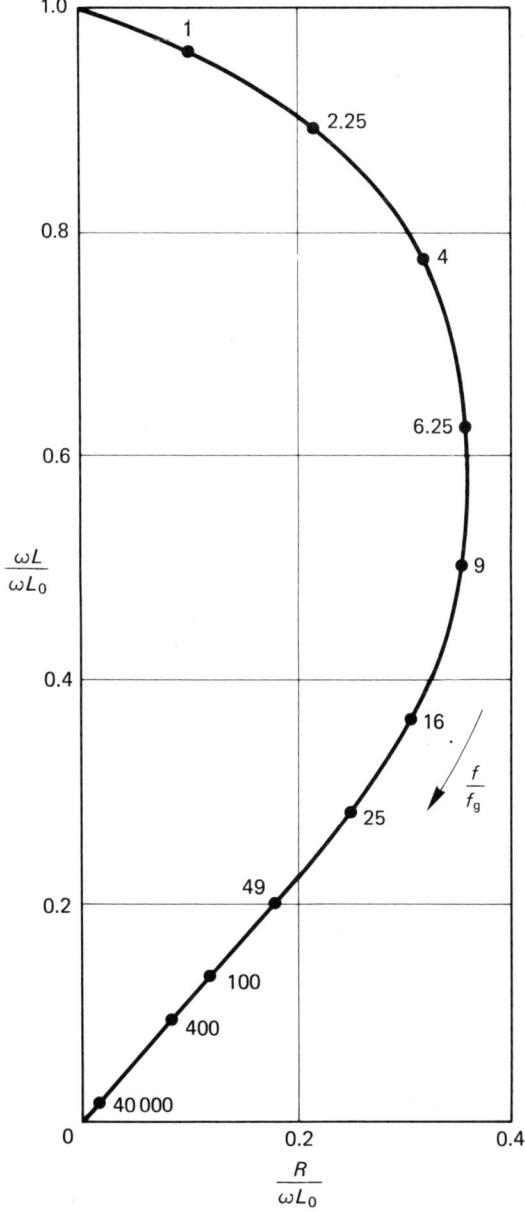

FIGURE 4.5   Normalised impedance plane diagram showing variation of resistance and reactance with frequency ratio.

From equation (4.1) it will be seen that the limit frequency, f_g, will be increased if the conductivity of the test specimen is decreased or, alternatively, if the specimen conductivity is increased the ratio f/f_g will also increase. If, in conjunction with a particular specimen the ratio f/f_g is equal to unity, then the coil impedance would be represented by point 1 on the graph (figure 4.5). If the conductivity of some other specimen is four times greater than that of the first sample and all other variables remain unchanged, then the coil impedance will be represented by point 4 on the same graph.

In figure 4.6 the variation of coil impedance with specimen conductivity is shown. In addition, the effects of changes in another variable, namely the relative size of specimen and coil, are also shown in this figure.

For a cylindrical specimen within a solenoid-type coil a *fill factor* can be determined. This is given by the ratio d^2/D^2, where d and D are the diameter of the specimen and the internal diameter of the coil respectively. The three curves in figure 4.6 represent three different values of 'fill factor', these being 1.0, 0.63 and 0.35, and indicate the effect variations in 'fill factor' have on impedance value. The dotted lines in figure 4.6 are lines for constant conductivity.

The set of curves in figure 4.7 shows the effect of variations in the relative magnetic permeability of magnetic materials on coil impedance, with the 'fill factor' remaining constant. Again the dotted lines in the figure are lines for constant conductivity.

Impedance plane diagrams of this type are useful since it becomes possible to separate variations in component diameter and conductivity at constant permeability or variations in permeability and conductivity at constant diameter.

Figure 4.6 indicates that at low frequency the diameter and conductivity graphs run almost in the same direction. At high frequencies, however, they form an angle of at least 45°. It is possible by suitable instrumentation to separate the signals obtained from these variations when a phase difference of ⩾45° exists. Hence, at a fixed frequency, it is possible to sort specimens with the same permeability according to their conductivity or diameter (see also section 4.13).

Figure 4.7 indicates how variations in permeability and conductivity at constant diameter can be interpreted independently for specimens with the same diameter. However, in this case a low frequency must be employed since the graphs subtend small angles at high frequencies.

4.6 Lift-off factor and edge effects

When an inspection coil is energised in air a recording instrument always gives an indication, even in the absence of a specimen. As the coil is moved closer to a conductor (component) the indication will change, and the magnitude of the change will increase until the coil is directly on the conductor. The changes in spacing between coil and conductor are termed *lift-off*. The lift-off effect is so pronounced that small variations in spacing can mask many indications resulting

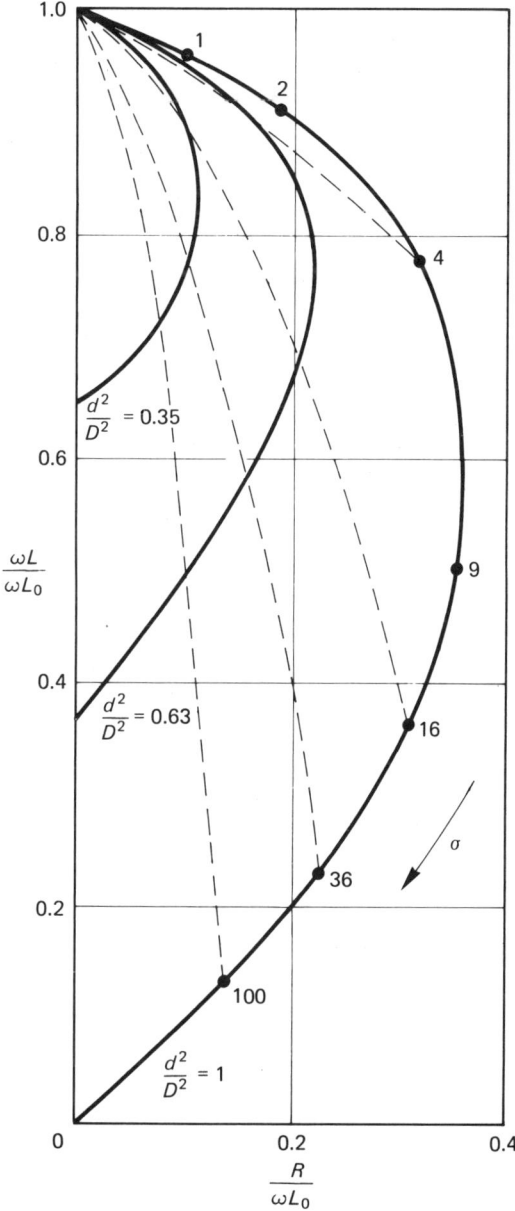

FIGURE 4.6   Variation of coil impedance with conductivity and relative size of specimen and coil.

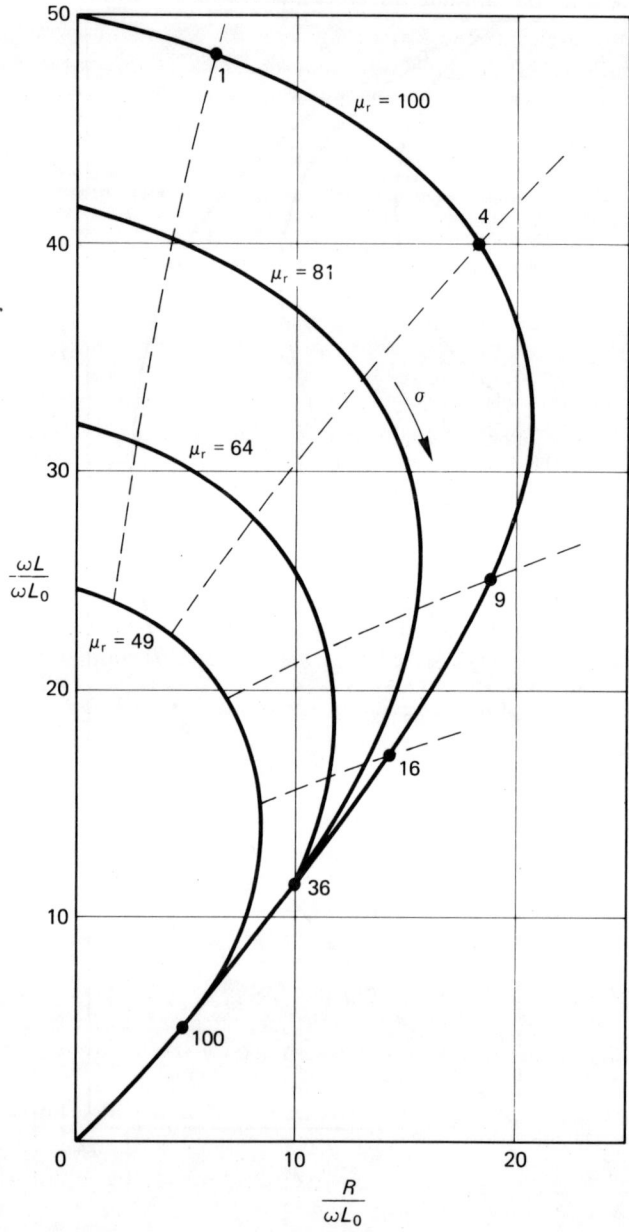

FIGURE 4.7   Effect of varying magnetic permeability on coil impedance.

from conditions of primary interest. The lift-off effect accounts for the extreme difficulty experienced in scanning components of complex shape.

At component edges, eddy current flow is distorted, because the eddy currents are unable to flow beyond this limiting barrier. The magnitude of this *edge effect* is usually very large, and hence, inspection is inadvisable close to edges. In general, it is recommended that inspection be limited to an approach of 3 mm to a component edge.

4.7 Skin effect

Eddy currents are not distributed uniformly throughout a part under inspection. They are most dense at the component surface, immediately beneath a coil, and become progressively less dense with increasing distance from the surface. At some distance below the surface of large components, eddy current flow is negligible. This phenomena is commonly termed 'skin effect'.

The eddy current density decreases exponentially with distance from the specimen surface, and the depth below the surface at which the magnitude of the eddy current is reduced to $1/e$ of the value of surface current (approximately 37 per cent) is termed the *standard penetration depth*, S. This depth can be estimated using the relationship

$$S = 15\,900 \sqrt{\left(\frac{\rho}{\mu_r f}\right)}$$

where S is the standard penetration depth (mm)
       ρ is the resistivity of the material (Ω mm)
      μ_r is the relative magnetic permeability ($\mu_r = 1$ for non-magnetic materials)
      f is the frequency (Hz).

The standard penetration depth for some materials, as a function of frequency, is shown in figure 4.8 and it will be seen that the standard penetration depth is not a constant for any particular material but it decreases as the test frequency increases.

When the thickness of a testpiece is less than approximately three times the standard penetration depth, the distribution pattern of the eddy currents will become distorted and the extent of such distortion will vary with the thickness of the material. This effect is shown schematically in figure 4.9.

It follows then, that for materials of thin section a change in thickness will alter the impedance value of a test coil. An eddy current system therefore, when calibrated against known standards, can be successfully used for the accurate measurement of the thickness of thin materials. Thickness measurements can

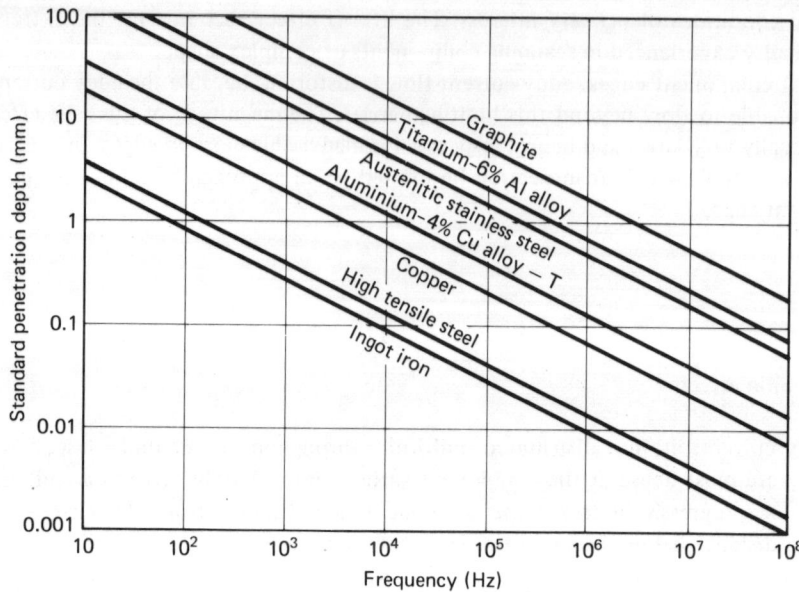

FIGURE 4.8   Standard penetration depth as a function of frequency for several
              materials.

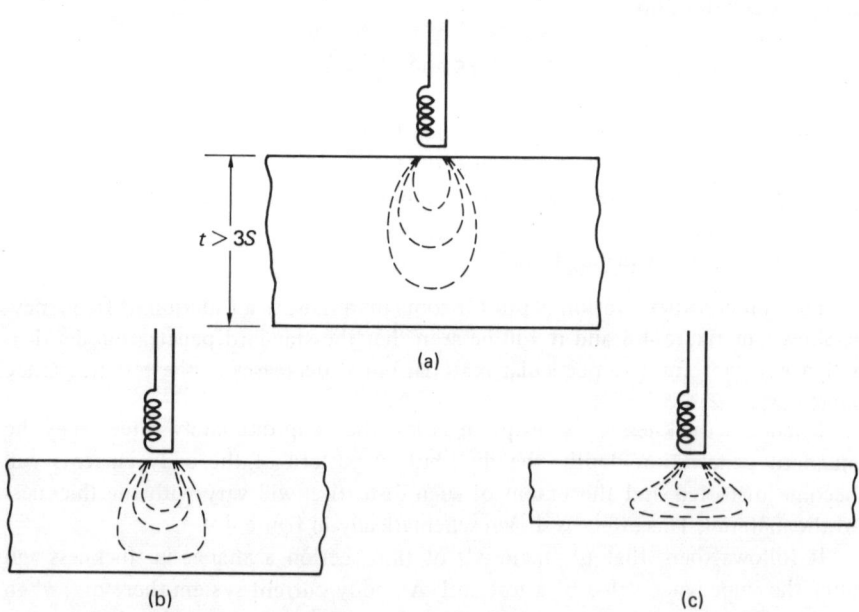

FIGURE 4.9   Schematic representation of distortion of eddy current distribu-
              tion in thin sections.

also be made using ultrasonic techniques (refer to chaper 5), but in this case the degree of accuracy possible diminishes when the material is very thin. The reverse is true for thickness measurements made using eddy current techniques and so the two different methods become complementary to each other.

4.8 Inspection frequency

The inspection frequencies used in electrical techniques range from 20 Hz to 10 MHz. Inspection of non-magnetic materials is performed at frequencies within the range 1 kHz–5 MHz, while frequencies lower than 1 kHz are often employed with magnetic materials. The actual frequency used in any specific test is normally a compromise in order to attain optimum sensitivity at the desired penetration depth.

It is worth noting that similar indications will be obtained from similar defects in a wide range of materials provided that the inspection frequencies are adjusted so that the ratio of inspection frequency to the limit frequency, f/f_g, has a constant value.

For non-magnetic materials the choice becomes relatively simple when it is required to detect surface flaws only. Frequencies as high as possible (several MHz) are used. However, detection of sub-surface flaws at a considerable depth demand low frequencies, sacrificing sensitivity. Hence, small flaws may not be detected under such conditions.

Inspection of ferro-magnetic materials demands very low frequencies because of the relatively low penetration depth in these materials. Higher frequencies can be used to inspect for surface conditions only. However, even the higher frequencies used in these applications are still considerably lower than those used to inspect non-magnetic materials for similar conditions.

4.9 Coil arrangements

There are several different coil arrangements which can be used in eddy current testing and some of the more common types are shown in figure 4.10.

A single primary solenoid type coil may be used for the routine inspection of cylindrical bars or tubes. The testpiece is passed through the coil (figure 4.10a). Variations in coil impedance value as the testpiece moves through the coil will indicate the presence of flaws. When it is not possible to place a coil around the outside of a tube as, for example, during the routine *in situ* inspection of heat exchanger or condenser tubes, the coil can be wound on a bobbin and inserted into the bore of the tubes. Frequently a double primary coil system is used in tube inspection (figure 4.10b). The two coils are identical and are connected to adjacent arms of a bridge network. When the tube is unform the bridge circuit will be balanced but if one coil is in proximity to a crack, or an area where corrosion has caused a thinning of the tube wall, the bridge will be thrown out of balance.

The differential coil system (figure 4.10c) is sometimes used instead of the double primary coil for tube and bar inspection. When the material is uniform there will be zero voltage across **AB** but if a flaw existed at some point X in the specimen, a voltage would exist between **AB**. One of the most widely used arrangements is the surface coil (figure 4.10d) in which a coil is wound around a ferrite

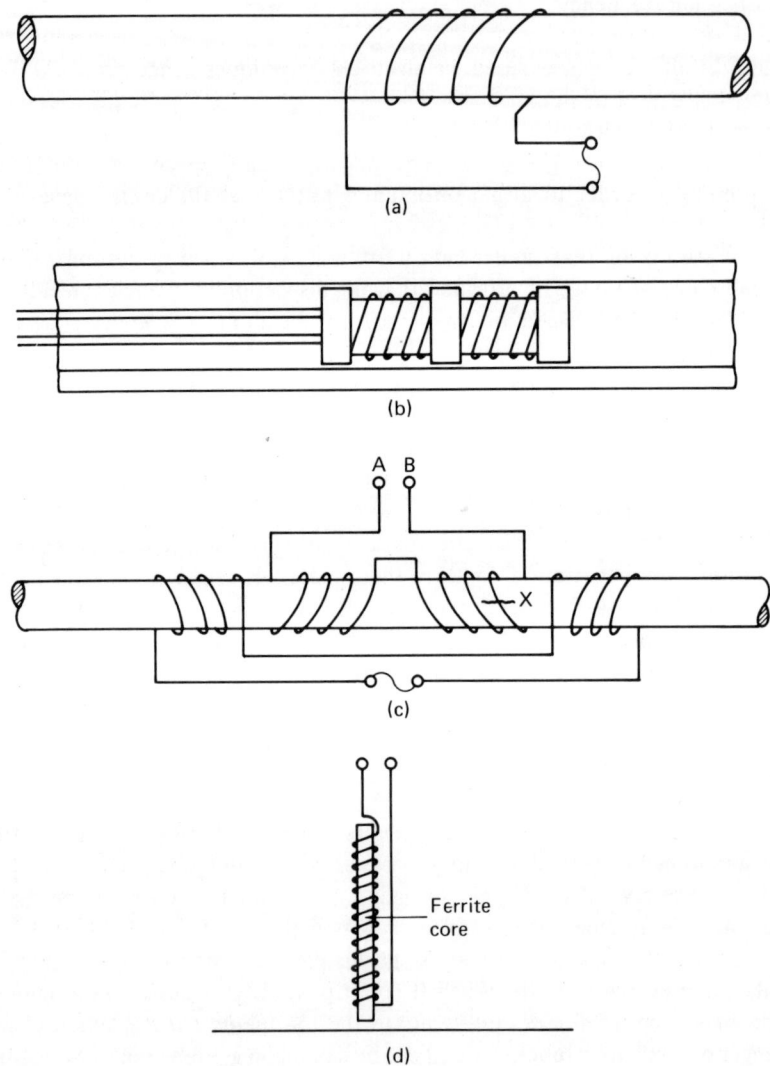

FIGURE 4.10   Coil arrangements: (a) single primary coil around bar; (b) twin-coil bobbin inside tube; (c) differential coil system; (d) surface coil with ferrite core.

core. The coil is held normal to the surface of the material being inspected. The ferrite core will concentrate the magnetic flux and increase the sensitivity for the detection of small defects.

4.10 Inspection probes

For many test and inspection purposes, a coil or coils are mounted in a holder as an inspection probe. As stated earlier, a coil is frequently wound around a ferrite core and, while the search coil is generally protected by a plastic casing, the end of the ferrite core often projects beyond the plastic case. Eddy current test probes do not need any coupling fluid between them and the testpiece, unlike ultrasonic probes, because they are coupled to the material by a magnetic field, and consequently little if any surface preparation is necessary prior to inspection. Many types of inspection probe have been designed but these can be broadly divided into surface probes and hole probes. The arrangement in two typical types of surface probe is shown in figure 4.11.

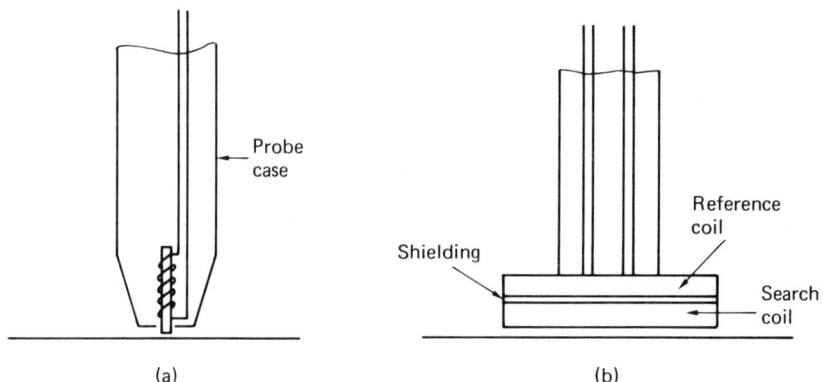

FIGURE 4.11   Surface probes: (a) for defect detection; (b) for coating thickness
measurement.

The single coil probe shown in figure 4.11a is ideal for the detection of surface defects such as small cracks. The probe should be held normal to the component surface otherwise incorrect indications could occur, and the probe may be held in a jig to ensure that it is always normal to the testpiece surface. The signal from this type of single coil probe would be connected to test equipment with a resonant circuit.

The twin coil probe type shown in figure 4.11b is suitable for the measurement of the thickness of surface coatings or for conductivity measurements. The signals from twin coil probes are best analysed using a test set with a bridge circuit.

Bore or hole probes possess single or twin coils and are designed to be inserted into tubes or holes, such as bolt-holes in plate. For the examination of tubing a twin-coil probe in which the axis of the coils is coincident with the tube axis will be suitable (see figure 4.10b). However, for the inspection of cylindrical bolt or rivet holes in sheet or plate material, a different arrangement is used. The most common type of defects which are sought in fastener holes are fine radial cracks. For the greatest sensitivity in the detection of this type of defect what is needed is, in effect, a surface probe coil configuration. To this end the search coil is usually a very small coil aligned radially but also offset from the centre of the hole (figure 4.12). During the inspection of a hole the probe is rotated through 360°. Not only will the presence of a crack be noted, but also its exact position at the circumference of the hole can be determined. In addition to the general types of surface and bore probes mentioned above, special probes can also be designed to satisfy the special particular requirements of a wide range of applications.

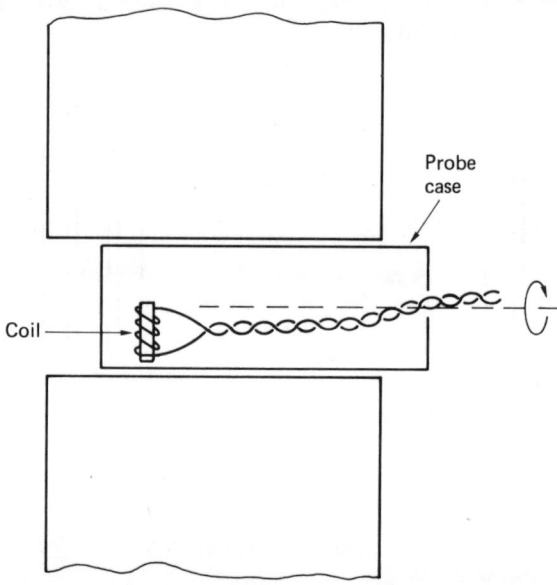

FIGURE 4.12   Arrangement of bore probe.

4.11 Types of circuit

The circuitry used in eddy current test equipment either involves a bridge network or utilises a resonant circuit. A bridge circuit, as shown in figure 4.13a, is very suitable when a twin-coil inspection probe is used, or when a single search coil is used in conjunction with a reference coil. In the former case, with an inspection probe containing two matched coils, the bridge will be in balance when the component being tested is defect-free but will be thrown out of balance when the probe is in proximity to a defect. Any voltage developed across BD in the bridge network can be recorded on a chart recorder to give a permanent record. When a twin-coil arrangement, as shown in figure 4.10b, is used for tube inspection, the bridge will be unbalanced in one direction as the first coil reaches a defect and then will be thrown out of balance in the reverse sense as the second coil reaches the defect (see figure 4.13b).

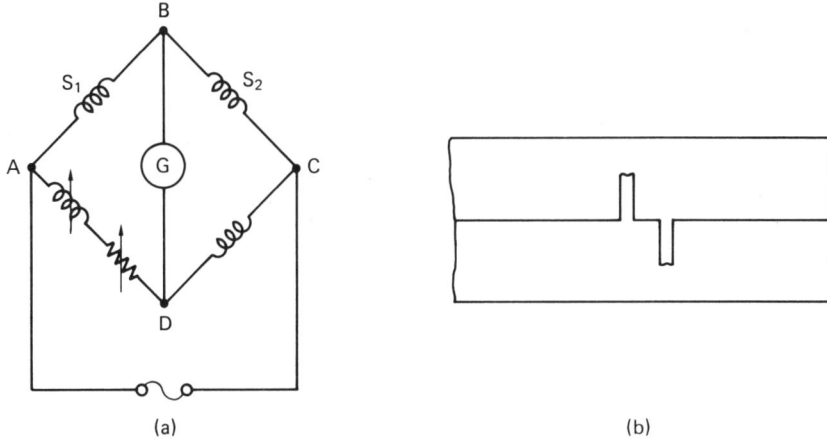

(a)   (b)

FIGURE 4.13   (a) Bridge network. S_1 and S_2 are either twin search coils or a search coil and a reference coil. (b) Section of recording trace produced by twin coil in a tube, showing a defect response.

Similarly, in the case of a single search coil in conjunction with a reference coil, the two coils are connected in adjacent arms of the bridge and the bridge will be unbalanced when the search coil is in proximity to a defect or because of lift-off effects.

Resonant circuits

A coil possesses a certain capacitance as well as inductance but the capacitance is usually small in relation to the inductance. If, however, a capacitor is connected in the same circuit as an inductor, the inductive reactance will increase with an increase in frequency and the capacitive reactance will decrease with an increase in frequency. Consequently, there will be some value of frequency, the *resonant frequency*, at which the two effects will be equal and opposite.

One major benefit of using an eddy current search probe in conjunction with a resonant circuit is that it is possible to suppress lift-off effects. Figure 4.14 shows a typical eddy current circuit in which the search probe is a parallel tuned circuit connected to an oscillator. When the impedance of the probe coil changes because of variations in the testpiece, so the oscillator frequency will change and will now differ from the frequency of the anode tuned circuit. This will result in a change of impedance in the coil of the anode tuned circuit and will give a change in the reading of the meter which is connected to the secondary windings of the anode coil. In operating, the search probe is placed on the metal testpiece and by adjustment of the lift-off control (a variable capacitor in the anode circuit) the anode circuit is tuned to the probe circuit to give a zero reading on the meter. Lifting the probe coil from the testpiece surface will give an impedance change which may be reflected in a reading on the meter, but by further fine adjustment of both the lift-off and the meter 'set zero' controls together it is possible to obtain a zero meter reading with the probe coil either on or off the testpiece surface, thus eliminating lift-off effect. Any meter reading will now indicate the presence of a defect in the material.

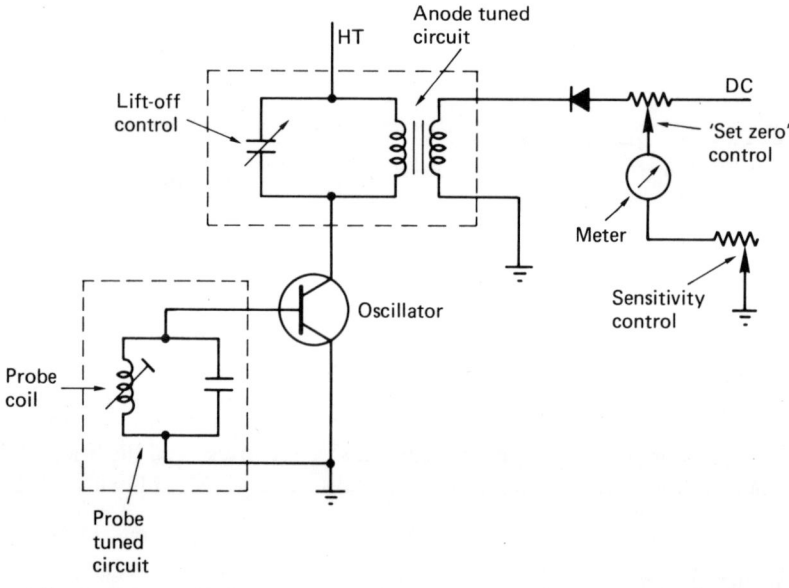

FIGURE 4.14   Typical tune resonant circuit for eddy current test equipment.

There are other types of resonant circuit which have been developed and used in test instruments to suppress lift-off effect.

4.12 Reference pieces

It is necessary to calibrate eddy current test equipment, and reference testpieces for calibration purposes should be made from material of similar type and quality to that which is to be tested so as to have the same conductivity value. A test-block should contain a series of defects of known size and shape and these are frequently made by making several fine saw-cuts of varying but known depth. However, this is not wholly satisfactory as a saw-cut will possess a specific width because some metal will have been removed by the cutting action, and the magnitude and phase of the change in coil impedance will not be the same as that produced by a naturally produced fatigue crack of the same depth. In many situations users will also use defective parts containing, for example, fatigue cracks as reference and calibration pieces.

4.13 Phase analysis

One method of representing the signals from eddy current inspection probes is by the phasor technique or phase analysis. When it is only necessary to detect changes in one of the parameters which affect impedance and all other factors are constant, then the measurement of a change in impedance value will reflect a change in the one parameter. However, there are many instances where it becomes necessary to separate the responses from more than one parameter, and to separate the reactive and resistive components of impedance. This requires the use of more sophisticated instruments but in this way it becomes possible to identify the type of defect present and not merely its position.

There is a phase difference between the reactive and resistive components of the measurement voltage. Consider the voltage as vectors A and B. The frequency is the same for both and, therefore, the radian velocity ω will be the same for both ($\omega = 2\pi f$).

The equations describing the vectors will be of the form $K(\sin \omega t + \phi)$ where ϕ is the phase angle. A plot of these vector equations will give sinusoidal curves, as in figure 4.15.

Figure 4.15a shows two vectors, A and B, both rotating with radian velocity ω and separated by an angle ϕ.

The sinusoidal curves corresponding to these vectors are shown in figure 4.15b and the amplitude of each sinusoid represents the magnitude of the vector.

The resistive and reactive components of a measurement (probe coil) voltage can be fed to the 'X' plates and 'Y' plates respectively of a cathode ray oscilloscope and displayed as a two-dimensional representation.

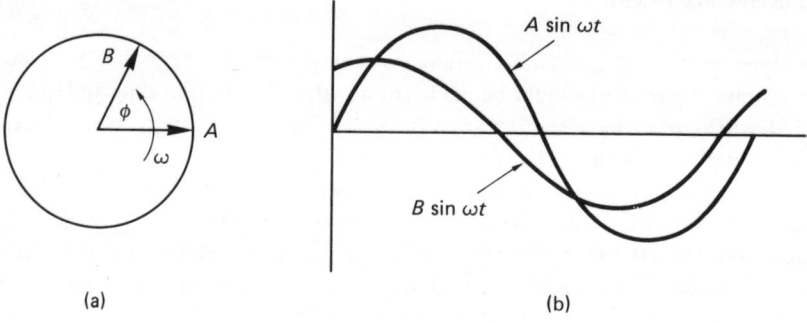

FIGURE 4.15　Phase differences: (a) two vectors separated by phase angle;
(b) sinusoidal curves with phase difference.

4.14 Display methods

There are several ways in which impedance data can be displayed on the screen of
a cathode ray oscilloscope.

One method is the *vector point* method. In this a spot is projected on to the
screen, representing the impedance, Z_0 (corresponding to the resistive and reactive
vectors). When the probe coil is placed in contact with a test specimen the spot
will move to correspond with the impedance change to Z_1 (refer to figure 4.4).
The position Z_1 for a reference test-block can be adjusted to be at the centre of
the screen. Any variation or anomaly in the component under test will cause a
movement of the spot and the direction of this movement will indicate the cause
of the variation. When more than one variable is present they can often be isolated
by vector analysis (see figure 4.16).

A storage type oscilloscope is used with eddy current test equipment.

The impedance changes caused by various types of defect or by changes in
conductivity will give screen displays as shown in figure 4.16b and c.

Ellipse method

This display method is used in conjunction with twin coil or differential coil
probes. With the probe coil in position with a reference block or ideal material,
the screen display is set as a horizontal line. A variation in one variable can be
denoted by a change in the angle of the line and a second variable can be denoted
by the formation of an ellipse. The values of both variables can be determined by
analysing both the position and shape of the ellipse (figure 4.17). It becomes
possible in many cases positively to identify the type of defect and its size. For
example, in the inspection of tubing using a differential coil probe it is possible to
identify cracks, holes, wall thinning caused by corrosion and also whether the
defect is on the inside or outside of the tube.

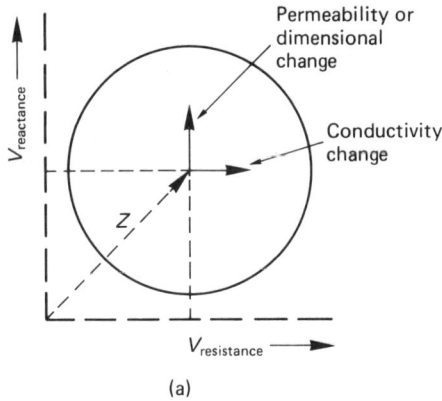

(a)

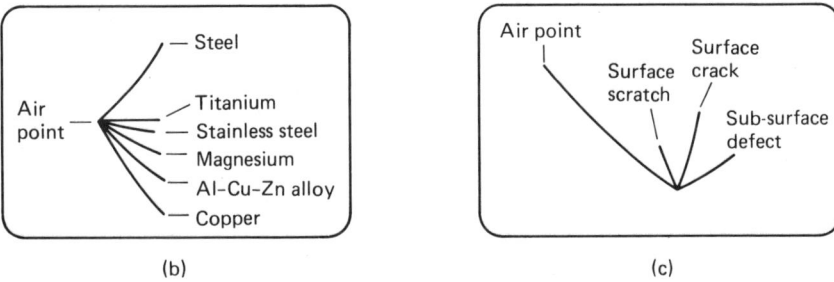

(b)   (c)

FIGURE 4.16   (a) Vector point. (b) Impedance plane display on oscilloscope, showing differing conductivities. (c) Impedance plane display, showing defect indications.

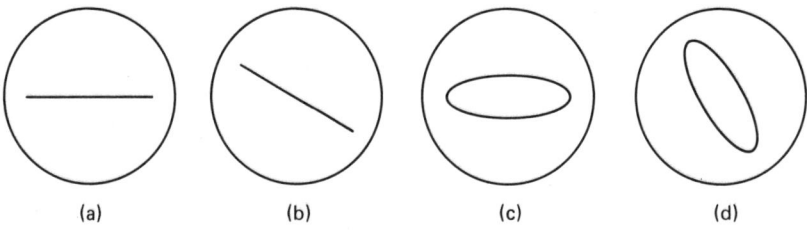

(a)   (b)   (c)   (d)

FIGURE 4.17   Ellipse method of phase analysis: (a) reference line; (b) change in one parameter, for example, dimensions; (c) change in one parameter, for example, conductivity; (d) change in both parameters.

4.15 Typical applications of eddy current techniques

The eddy current equipment which is used for the quality control inspection of the products of many material manufacturing processes is often completely automatic and highly sophisticated. Tube, bar and wire are frequently checked in this way and a complete installation for the inspection of bar material is illustrated in figure 4.18. This installation comprises a debundling table, from which the bars travel on a conveyor (on the right of the picture) and through an eddy current test head. Any defect indications are displayed on the screen of the cathode ray tube of the Defectometer and also recorded on tape. Immediately to the left of the test head is a paint spray head and any defective portion of bar will be marked at this point. The bar then passes through a demagnetising head and into an accept/reject mechanism which separates the bars into acceptable and reject containers. The speed at which bars pass through the test equipment is 2 m/s.

FIGURE 4.18   Eddy current inspection system for rolled bars (courtesy of Wells–Krautkramer Ltd).

Figure 4.19 illustrates the use of an eddy current test system for the routine inspection of aircraft undercarriage wheels. The wheel is placed on a turntable and the probe coil which is mounted at the end of an adjustable arm, is positioned near the bottom of the wheel. As the wheel turns on the turntable so the probe arm moves slowly up the wheel, giving a close helical search pattern. It is necessary to use a second probe to check under the wheel flange. A hand-held probe is used for this part of the inspection, as shown in figure 4.19b.

The eddy current system is a highly versatile system and can be used to detect not only cracks but several other conditions, including corrosion. Corrosion of hidden surfaces as, for example, within aircraft structures, can be detected using phase-sensitive equipment. It is a comparative technique in that readings made in a suspect area are compared with instrument readings obtained from sound, non-corroded material.

Eddy current equipment can be used for the identification and sorting of materials. Phase-sensing equipment is necessary for this purpose, particularly if it is required to differentiate between materials which are close to one another in composition (see also figure 4.16b).

The ability of eddy current techniques to determine the conductivity of a material has been utilised for the purpose of checking areas of heat-damaged skin on aircraft structures. If the type of aluminium alloy used in aircraft construction becomes over-heated it could suffer a serious loss of strength. This is accompanied by an increase in the electrical conductivity of the alloy. The conductivity of sound material is generally within the range of 31 to 35 per cent IACS. Defective or heat-damaged material would show a conductivity in excess of 35 per cent IACS.

Eddy current equipment may also be used for measuring the thickness of either conducting or non-conducting coatings on ferrous or non-ferrous base materials, but the measurement may be difficult if the conductivities of the coating material and basis metal are similar to one another. Test equipment specifically designed for coating thickness measurement of thin coatings (coating thickness less than 0.15 mm) is available from several manufacturers.

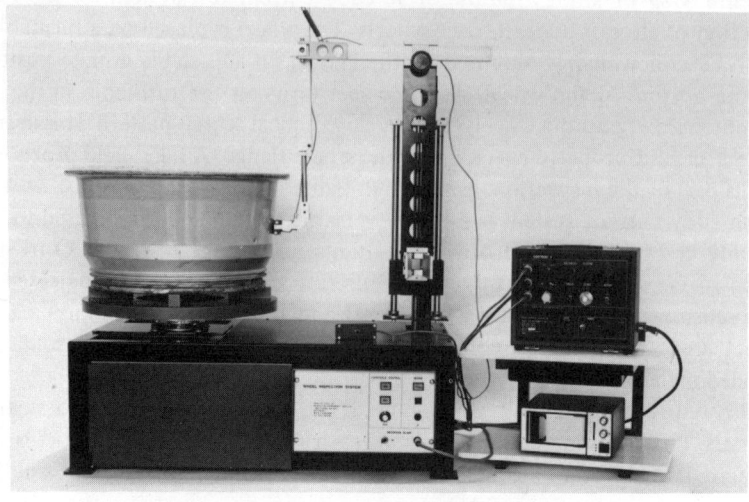

(a)

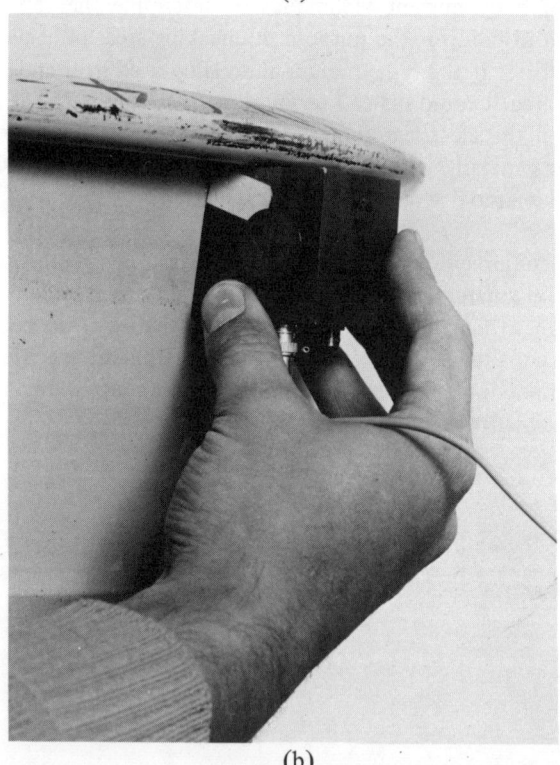

(b)

FIGURE 4.19 Aircraft wheel inspection system: (a) complete installation, showing probe in contact with wheel on turntable; (b) hand-held probe
for wheel flange inspection (courtesy of Inspection Instruments
Ltd).

5

Ultrasonic Testing

5.1 Introduction

Ultrasonic techniques are very widely used for the detection of internal defects in materials, but they can also be used for the detection of small surface cracks. Ultrasonics are used for the quality control inspection of part processed material, such as rolled slabs, as well as for the inspection of finished components. The techniques are also in regular use for the in-service testing of parts and assemblies.

Sound is propagated through solid media in several ways and the nature of sound will be considered first.

5.2 Nature of sound

Sound waves are elastic waves which can be transmitted through both fluid and solid media. The audible range of frequency is from about 20 Hz to about 20 kHz but it is possible to produce elastic waves of the same nature as sound at frequencies up to 500 MHz. Elastic waves with frequencies higher than the audio range are described as *ultrasonic*. The waves used for the non-destructive inspection of materials are usually within the frequency range 0.5 MHz to 20 MHz.

In fluids, sound waves are of the longitudinal compression type in which particle displacement is in the direction of wave propagation; but in solids, they are shear waves, with particle displacement normal to the direction of wave travel, and elastic surface waves can also occur. These latter are termed *Rayleigh waves*.

5.3 Wave velocity

The velocity of longitudinal compression waves, V_c, in a fluid is given by

$$V_c = \left(\frac{K_a}{\rho}\right)^{1/2}$$

where K_a is the adiabatic volume elasticity and ρ is the density. In solids, the velocity of compression waves is given by the expression

$$V_c = \left(\frac{E(1-v)}{(1+v)(1-2v)\rho}\right)^{1/2}$$

where E is the modulus of elasticity and v is Poisson's ratio. The velocities of longitudinal waves in several media are given in table 5.1.

Table 5.1   Velocities of longitudinal sound waves

Medium	Velocity V_c (m/s)
Air*	330
Water	1430
Oil	1740
Aluminium	6190
Copper	4600
Magnesium	5770
Steel	5810
Perspex	2730
Polyethylene	2340

*At atmospheric pressure and 15°C.

The velocity of shear waves, V_s, within a solid is roughly half the longitudinal wave velocity and is given by the expression

$$V_s = \left(\frac{G}{\rho}\right)^{1/2}$$

where G is the modulus of rigidity of the material.

The velocity of Rayleigh waves, V_r, in a solid is about 90 per cent of that of shear waves, and is given by

$$\frac{V_r}{V_s} = \frac{0.87 + 1.12v}{(1+v)}$$

5.4 Wavelengths

The wavelength, λ, is related to the frequency and wave velocity as: $V = \lambda f$ where f is the frequency. Table 5.2 gives the wavelengths of sound in various materials at several frequencies.

In common with other types of wave, sound waves will only be reflected effectively by objects which have dimensions equal to or greater than the wavelength of the radiation. It will be seen from table 5.2 that an ultrasonic beam with a frequency of 10 MHz will be capable of detecting defects in steel of sizes greater than 0.58 mm, but only defects larger than 4.65 mm would be observed if a frequency of 1.25 MHz were used.

Table 5.2 Wavelengths of sound (compression waves) in some materials

Material	λ (mm) for frequency (MHz) of			
	1.25	2.5	5.0	10.0
Air	0.26	0.13	0.066	0.033
Water	1.14	0.57	0.286	0.143
Oil	1.39	0.70	0.35	0.175
Aluminium	4.95	2.48	1.24	0.62
Copper	3.68	1.84	0.92	0.46
Magnesium	4.62	2.31	1.16	0.58
Steel	4.65	2.32	1.16	0.58
Perspex	2.18	1.09	0.55	–
Polyethylene	1.87	0.94	0.47	–

5.5 Generation of ultrasound

Certain crystalline materials show the piezo-electric effect, namely, the crystal will dilate or strain if a voltage is applied across the crystal faces. Conversely, an electrical field will be created in such a crystal if it is subjected to a mechanical strain, and the voltage produced will be proportional to the amount of strain. Piezo-electric materials form the basis of electro-mechanical transducers. The original piezo-electric material used was natural quartz. Quartz is still used to some extent but other materials, including barium titanate, lead metaniobate and lead zirconate, are used widely. When an alternating voltage is applied across the thickness of a disc of piezo-electric material, the disc will contract and expand, and in so doing will generate a compression wave normal to the disc in the surrounding medium. When quartz is used the disc is cut in a particular direction from a natural crystal but the transducer discs made from ceramic materials such as barium titanate are composed of many small crystals fused together, the crystals being permanently polarised to vibrate in one plane only.

Wave generation is most efficient when the transducer crystal vibrates at its natural frequency, and this is determined by the dimensions and elastic constants of the material used. Hence, a 10 MHz crystal will be thinner than a 5 MHz crystal. A transducer for sound generation will also detect sound. An ultrasonic wave incident on a crystal will cause it to vibrate, producing an alternating current across the crystal faces. In some ultrasonic testing techniques two transducers are used — one to transmit the beam and the other acting as the receiver — but in very many cases only one transducer is necessary. This acts as both transmitter and receiver. Ultrasound is transmitted as a series of pulses of extremely short duration and during the time interval between transmissions the crystal can detect reflected signals.

5.6 Characteristics of an ultrasonic beam

The ultrasonic waves generated by a disc-shaped crystal will emerge initially as a parallel-sided beam which later diverges, as shown in figure 5.1. The spread of the beam, α, is related to the frequency and the disc dimensions by the relationship

$$\sin \frac{\alpha}{2} = \frac{1.12\lambda}{d}$$

where λ is the wavelength and d is the diameter of the disc (both in mm). For example, the angle of spread in aluminium for a beam of frequency 5 MHz generated by a crystal of 20 mm diameter is $8.5°$, but the angle is reduced to $5.5°$ for a 15 mm diameter crystal vibrating at 10 MHz. If the ultrasonic frequency is decreased so the wavelength approaches the dimensions of the crystal disc, waves are generated in all directions and there is no beam propagation.

An ultrasonic beam can be divided into three zones, the dead zone, the near zone and the far zone.

The dead zone   This is the distance below the surface of a material in which a defect cannot be detected. A crystal is stimulated into vibration by an exciting voltage for a very short time to produce a short duration pulse of ultrasound. The crystal, even though heavily damped, does not stop vibrating immediately on cessation of the exciting voltage but 'rings' for a short time. It is not possible to detect a flaw during this ringing time (see figure 5.1b). If the crystal is mounted in a suitably dimensioned block of perspex, the dead zone can be wholly contained within the probe block. However, signals from defects close to the surface may still be lost in the interface echo.

The near zone   The near zone is the zone in which the beam is almost parallel sided. The length l, of the near zone is given by the approximate relationship

$$l = \frac{d^2}{4\lambda}$$

where d is the crystal diameter. The detection sensitivity is not constant throughout the near zone and is greatest towards the far end of this zone.

The far zone   The far zone is the region beyond the near zone where beam spread occurs, and within this zone the sensitivity decreases with the square of the distance from the crystal.

Effects of probe diameter and wave frequency

As the diameter of a crystal is reduced then, for a constant frequency, the beam spread angle increases, the length of the near zone decreases and a lower intensity beam is generated.

An increase in wave frequency will have the effects of giving a larger near zone and a smaller beam spread. It will also give a better resolution of defects, but the beam will have a lower penetrating power.

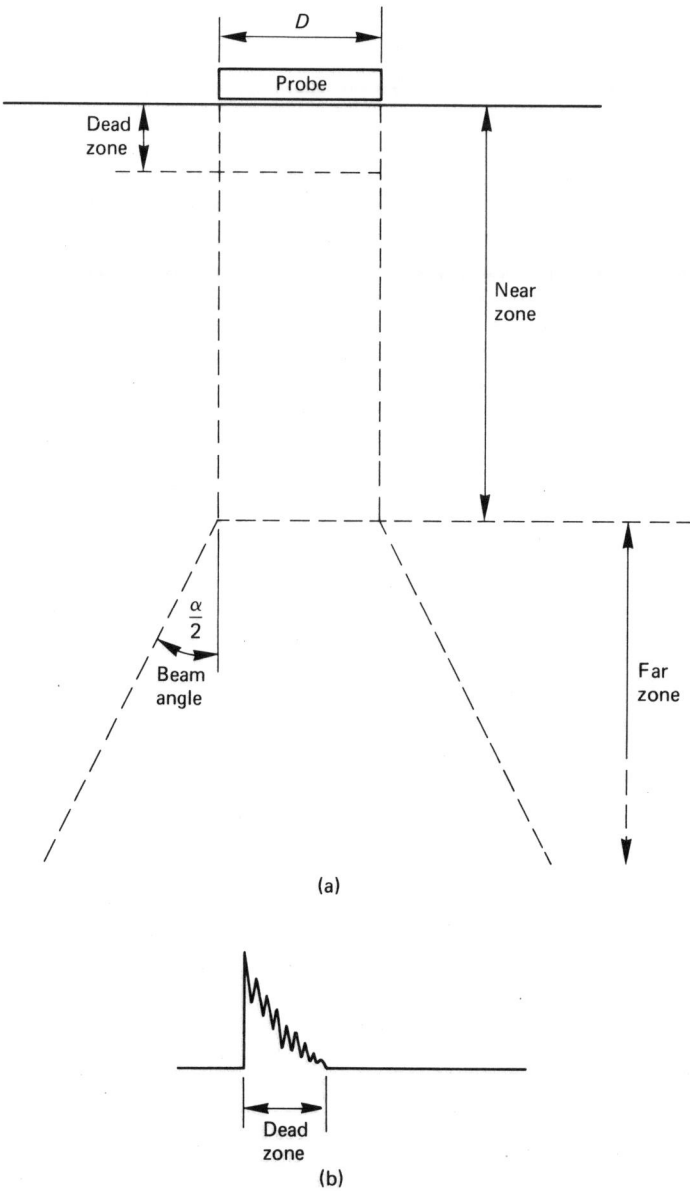

FIGURE 5.1   (a) Ultrasonic beam shape. (b) Appearance of dead zone on screen display.

5.7 Sound waves at interfaces

Reflection and transmission

There are limits for sound propagation at an interface between two media of differing elastic properties. When a beam of longitudinal compression sound waves reaches a boundary between two media, a proportion of the incident waves will be reflected at the interface and a proportion will be transmitted across the interface. For normal incidence waves the transmission across the interface will be of the compression wave type. The reflection coefficient, R, and the transmission coefficient, T, can be determined from the acoustic impedances, Z_1 and Z_2, of the two media in contact. The acoustic impedance, Z, of a material is given by: $Z = \rho V_c$ where ρ is the density of the material and V_c is the velocity of a compression wave in the medium.

The reflection coefficient, R, for sound waves travelling through medium 1 at an interface between medium 1 and medium 2 is given by

$$R = \frac{Z_2 - Z_1}{Z_2 + Z_1} = \frac{\rho_2 V_{c2} - \rho_1 V_{c1}}{\rho_2 V_{c2} + \rho_1 V_{c1}}$$

The corresponding transmission coefficient, T, is given by

$$T = \frac{2Z_1}{Z_2 + Z_1} = \frac{2\rho_1 V_{c1}}{\rho_2 V_{c2} + \rho_1 V_{c1}}$$

The acoustic impedances of several different media are given in table 5.3.

Table 5.3   Acoustic impedances of some media

Medium	Acoustic impedance (MPa s/m)
Air	4.04×10^{-4}
Water	1.43
Oil	1.58
Aluminium	16.77
Copper	41.08
Magnesium	10.04
Steel	45.72
Perspex	3.22
Polyethylene	2.16

Reflection of sound at an air/metal interface is practically 100 per cent at the frequencies commonly used in ultrasonic testing and, therefore, sound cannot be transmitted easily into a metal across an air gap. When a fluid such as oil or water is employed as a coupling agent between a transducer crystal and a metal, the reflection coefficient is reduced. For example, approximately 94 per cent reflection takes place at an oil/steel interface, or in other words, about 6 per cent of the incident sound energy will be transmitted across the interface.

The efficiency of transmission will be improved considerably if the thickness of the fluid coupling between transducer and metal is only a very small fraction of the wavelength of sound in the medium. For example, the transmission factor may be close to the theoretical value of 13 per cent in the case of a perspex-faced crystal probe used in conjunction with a steel component provided the thickness of the couplant film is small. However, the transmission factor varies with film thickness and, unless a steady pressure is maintained between the transducer and the metal surface, the intensity of the ultrasonic beam entering the metal may be subject to considerable fluctuation.

Reflection and refraction

When the incident beam is at some angle other than normal, that portion of the beam which is transmitted across the interface will be refracted. However, there may be two refracted beams transmitted into the metal, because part of the transmitted energy is converted into the shear wave mode. One refracted beam will be of the compression type while the other will be a shear wave, as shown in figure 5.2. *Note:* at a solid/solid interface there will also be a reflected shear wave.

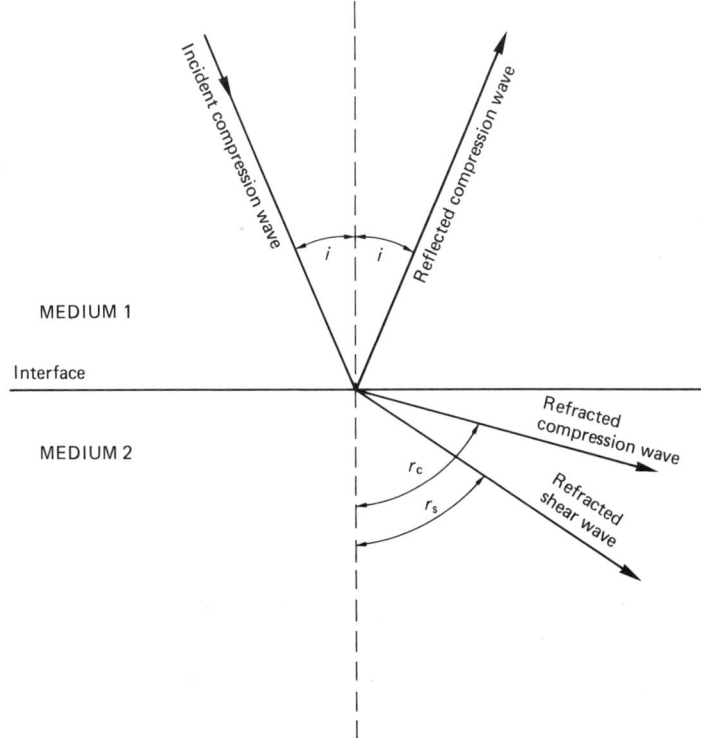

FIGURE 5.2   Reflection and refraction at an interface.

The directions of propagation of the refracted compression and shear wave components at the interface between two media may be determined using Snell's refraction law:

$$\text{Compression wave component} \quad \frac{\sin i}{\sin r_c} = \frac{V_{c1}}{V_{c2}}$$

$$\text{Shear wave component} \quad \frac{\sin i}{\sin r_s} = \frac{V_{c1}}{V_{s2}}$$

In ultrasonic inspection the presence of two types of wave with differing velocities within the material under test would give confusing results and so the angle of incidence is adjusted to be greater than the critical angle, i', for compression wave refraction, thus allowing only the refracted shear wave to be transmitted into the material. The critical angle for compression wave refraction is given by

$$i' = \sin^{-1} \frac{V_{c1}}{V_{c2}}$$

In an angle probe the transducer crystal is mounted in a block of perspex (see section 5.10). The critical angle of incidence for total reflection of the incident beam at a perspex/steel interface is $27.5°$. Angle wave probes for shear wave propagation are generally available with the following angles (all angles being measured from the normal): $35°$, $45°$, $60°$, $70°$ and $80°$. If it is required to generate a surface wave the angle of incidence should be adjusted to a second critical angle, i'', to give a Rayleigh wave at a refracted angle of $90°$. The critical angle for the generation of Rayleigh waves is given by

$$i'' = \sin^{-1} \left(\frac{V_{c1}}{V_{s2}} \right)$$

The value of this critical angle for a perspex/steel interface is $57°$. As will be seen later, surface waves are utilised for some types of ultrasonic inspection.

An ultrasonic beam being transmitted through a metal will be totally reflected at the far surface of the material, a metal/air interface. It will also be wholly or partially reflected by any internal surface, namely cracks or laminations, porosity and non-metallic inclusions, subject to the limitation mentioned in section 5.4 that the size of the object is not less than one wavelength. From this it follows that the sensitivity and defect resolution will increase as the frequency of the beam is increased.

Some metallic materials can only be inspected satisfactorily with a relatively low-frequency sound beam, because the use of high frequencies could cause a mass of reflections from a normal structural constituent which would mask the indications from defects. This situation arises in the inspection of grey cast iron components where the flakes or nodules of graphite in the iron structure may have a size of several millimetres.

5.8 Sound attenuation

A sound wave propagating through a material continually loses a part of its energy because of scattering at microscopic interfaces and also internal friction effects within the material. Grain boundaries, second phase particles and inclusions constitute microscopic interfaces in metals and alloys. The energy loss as the wave travels through the medium is termed *attentuation* and such loss occurs along the whole travel path of the sound beam. Attenuation losses, together with beam divergence, account for the major limitation imposed on the depth of penetration which may be achieved by sound waves during component inspection.

The extent of sound attenuation increases with an increase in frequency. The absorption of energy due to internal friction effects increases with the square of the frequency, but energy loss due to scatter at microscopic interfaces will increase according to some higher exponent of the frequency.

Energy losses/metre $= k_1 f^2 + k_2 f^x$ where f is the frequency, k_1 and k_2 are constants, and x has some value >2.

It is not possible to be more precise about the rate of energy loss due to scatter as there is such a wide variety of microscopic interfaces possible in metals and alloys. The frequency selected for the inspection of some particular material or component is frequently an optimum solution, and sometimes resolution is sacrificed in order to reduce attenuation and, so, achieve the required depth of penetration.

For the inspection of many metallic components, frequencies within the range of 10 to 20 MHz may be employed without incurring a serious loss of sound beam amplitude. However, the optimum range of frequencies used for the inspection of polymeric components is from 2 to 5 MHz because of the greater attenuation effects in these materials.

5.9 Display systems

In most ultrasonic test equipment the signals are displayed on the screen of a cathode ray oscilloscope. The basic block diagram for a typical flaw detector is shown in figure 5.3.

The functions of the various components in the system are as follows.

(a) A master timer controls the rate of generation of ultrasonic pulses, that is, the pulse repetition frequency (PRF).
(b) The pulse generator controls the amplitude of the pulses produced.
(c) The transmitter probe converts the electrical impulse into a mechanical vibration at the selected frequency. The transmitter probe may also act as the receiver for reflected echoes, or a separate receiver probe may be used. This depends on the type of inspection method being used.

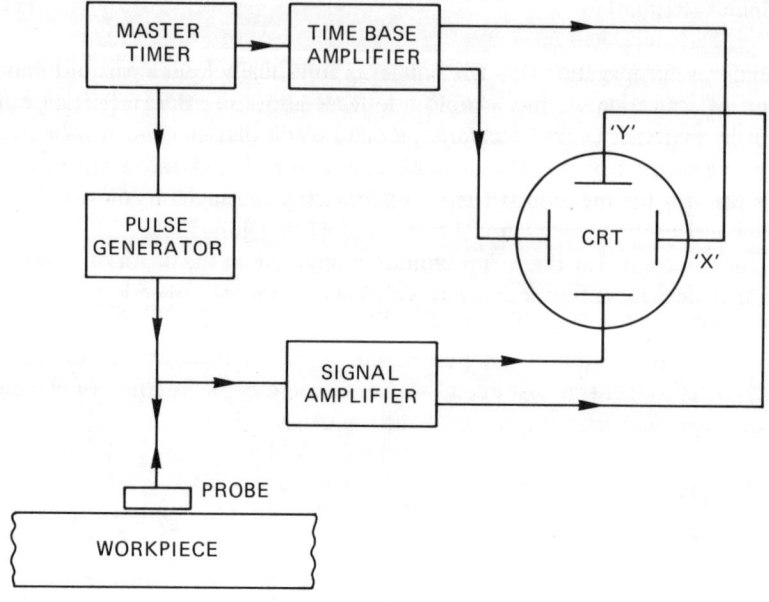

FIGURE 5.3   Block diagram for an ultrasonic flaw detector.

(d) The time base amplifier controls the rate of sweep of the electron beam across the face of the cathode ray tube. It is generally possible to vary the length of the time base between very wide limits so that one sweep of the screen may cater for any thickness of material from a few millimetres to several metres.

(e) The signal amplifier, as its name implies, amplifies the returned echo signal and feed this to the 'Y' plates of the cathode ray tube.

(f) The majority of ultrasonic sets are fitted with an attenutator. By means of this device the received signal strength can be measured in decibels relative to the signal from a standard reference.

5.10 Probe construction

There are several types of transmitter probe in use but each type consists of a crystal which is placed in contact, either directly or through a protective cover, with the material under test. There are several materials which may be used as transducer crystals and these include natural quartz, barium titanate, lead niobate and lithium sulphate. A step voltage, of short duration, is applied to the crystal and this causes the crystal to vibrate at its natural frequency. After the step voltage has been removed the crystal oscillation is required to die as soon as possible, and the crystal is usually backed by a damping material to assist this process.

Probes may be of the normal type, or be angled.

Normal probes

A normal probe is designed to transmit a compression wave into the test material at right angles to the material surface. In some cases the crystal surface is uncovered so that it may be placed directly, via an oil or water film, in contact with the test material. Alternatively, the crystal may be protected by a layer of metal, ceramic or perspex. In the last case the perspex block may be shaped to allow for normal transmission into material with a curved surface (see figure 5.4a)

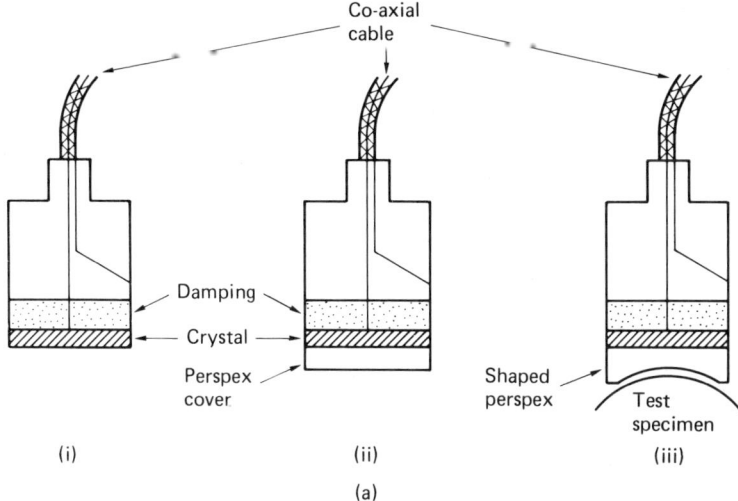

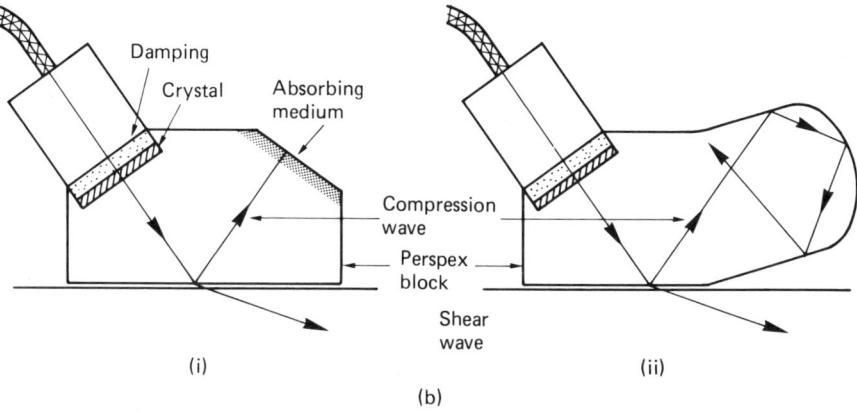

FIGURE 5.4   Probe construction. (a) Normal probes: (i) uncovered probe; (ii) covered probe; (iii) normal probe for curved surface. (b) Angle probes: (i) with sound absorbent; (ii) shaped for reflected wave dissipation.

Angle probes

Angle probes are designed to transmit shear waves or Rayleigh waves into the test material. The general construction of an angle probe is similar to that of a normal probe with the crystal embedded in a shaped perspex block. There is a reflected compression wave produced at the perspex/metal interface. This reflected wave could possibly return to the crystal and give confusing signals. To obviate this an absorbent medium, such as rubber, is built into the probe. An alternative method is to shape the perspex block in such a way that the reflected wave is 'bounced' around several times until its energy is dissipated. This is possible since perspex has a high absorption coefficient (see figure 5.4b).

5.11 Type of display

The information obtained during an ultrasonic test can be displayed in several ways.

'A' scan display

The most commonly used system is the 'A' scan display (see figure 5.5). A blip appears on the CRT screen at the left-hand side, corresponding to the initial pulse, and further blips appear on the time base, corresponding to any signal echoes received. The height of the echo is generally proportional to the size of the reflecting surface but it is affected by the distance travelled by the signal and attentuation effects within the material. The linear postion of the echo is proportional to the distance of the reflecting surface from the probe, assuming a linear time base. This is the normal type of display for hand probe inspection techniques.

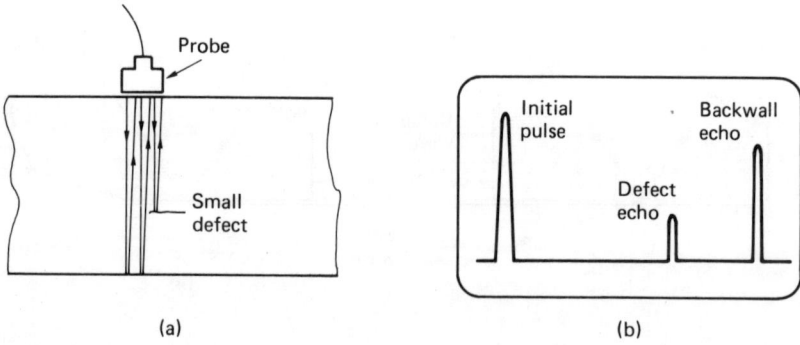

(a)   (b)

FIGURE 5.5   'A' scan display: (a) reflections obtained from defect and back-wall; (b) representation of 'A' scan screen display.

A disadvantage of 'A' scan is that there may be no permanent record, unless a photograph is taken of the screen image, although the more sophisticated modern equipment has the facility for digital recording.

'B' scan display

The 'B' scan enables a record to be made of the position of defects within a material. The system is illustrated in figure 5.6. There needs to be co-ordination between the probe position and the trace, and the use of 'B' scan is confined to automatic and semi-automatic testing techniques. With the probe in position '1' the indication on the screen is as shown in figure 5.6, with (i) representing the initial signal and (ii) representing the backwall. When the probe is moved to position '2', line (iii) on the display represents the defect. This representation of the testpiece cross-section may be recorded on a paper chart, photographed, or viewed on a long-persistence screen.

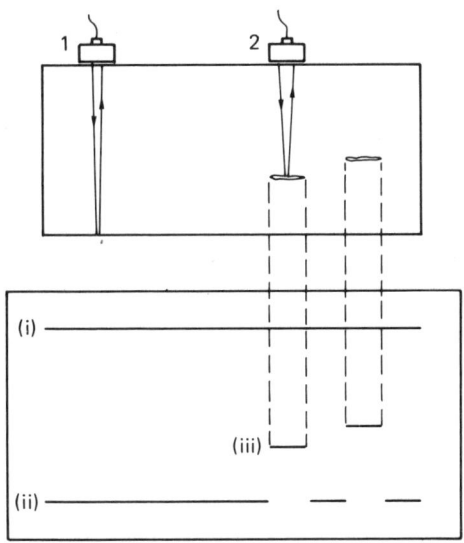

FIGURE 5.6   'B' scan display.

'C' scan display

While 'B' scan gives a representation of a side elevation of the testpiece, another method, termed 'C' scan can be used to produce a plan view. Again, 'C' scan display is confined to automatic testing.

5.12 Inspection techniques

The presence of a defect within a material may be found using ultrasonics with either a transmission technique or a reflection technique.

Normal probe reflection method

This is the most commonly used technique in ultrasonic testing, and is illustrated in figure 5.5. The pulse is wholly or partially reflected by any defect in the material and received by the single probe, which combines as transmitter and receiver. The time interval between transmission of pulse and reception of echo is used to indicate the distance of the defect from the probe.

The reflection method has certain advantages over the transmission method. These are:

(a) The specimen may be of any shape.
(b) Access to only one side of the testpiece is required.
(c) Only one coupling point exists, thus minimising error.
(d) The distance of the defects from the probe can be measured.

Normal probe transmission method

In this method a transmitter probe is placed in contact with the testpiece surface, using a liquid coupler, and a receiving probe is placed on the opposite side of the material (see figure 5.7).

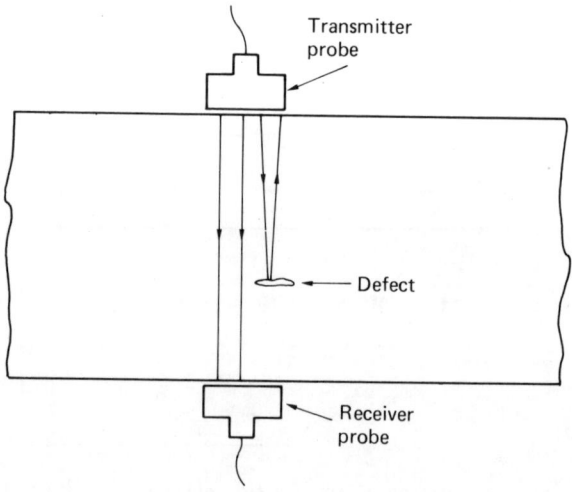

FIGURE 5.7   Normal probe transmission technique.

If there is no defect within the material, a certain strength of signal will reach the receiver. If a defect is present between the transmitter and receiver, there will be a reduction in the strength of the received signal because of partial reflection of the pulse by the defect. Thus, the presence of a defect can be inferred.

This method possesses a number of disadvantages. These are:

(a) The specimen must have parallel sides and it must be possible to reach both sides of the piece.
(b) Two probes are required, thus doubling the possibility of having inefficient fluid coupling.
(c) Care must be taken that the two probes are exactly opposite one another.
(d) There is no indication of the depth of a defect.

Angle probe transmission method

There are certain testing situations in which it is not possible to place a normal probe at right angles to a defect and the only reasonable solution is offered by angle probes. A good example of this technique is in the inspection of butt welds in parallel-sided plates. The transmitter and receiver probes are arranged as in figure 5.8a. If there is any defect in the weld zone, this will cause a reduction in the received signal strength. Distance **AB** is known as the 'skip distance' and for the complete scanning of a weld the probes should be moved over the plate surface as shown in figure 5.8b. In practice, both probes would be mounted in a jig so that they are always at the correct separation distance.

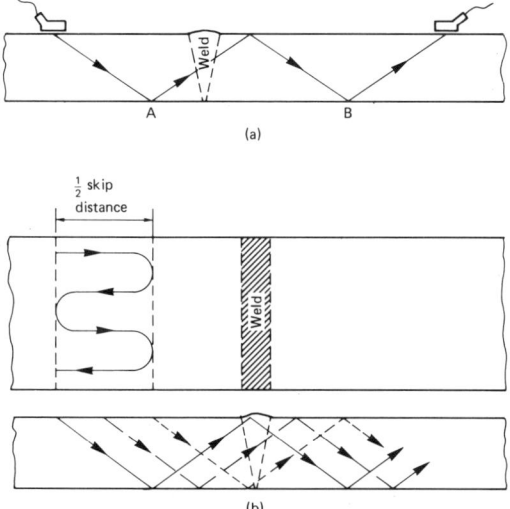

(a)

(b)

FIGURE 5.8   Angle probe transmission method: (a) probe positions and skip distance; (b) scanning method for complete inspection of butt weld.

Angle probe reflection method

Defects can also be detected using one angle probe in the reflection mode, as
shown in figure 5.9. It is important when using an angle probe in this type of test
that the flaw detector be accurately calibrated using a reference test-block. The
design and use of a calibration block are covered in a later section.

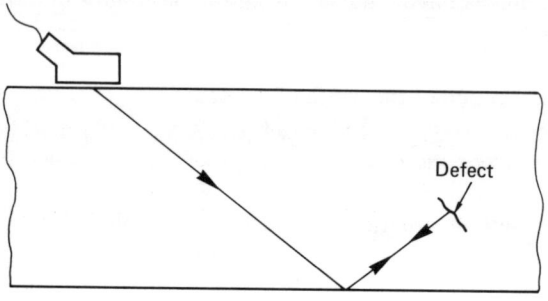

FIGURE 5.9   Reflective technique with angle probe.

Inspection using a surface wave probe

A Rayleigh, or surface, wave can be used for the detection of surface cracks (see
figure 5.10). The presence of a surface defect will reflect the surface wave to give
an echo signal in the usual way. Surface waves will follow the surface contours
and so the method is suitable for shaped components such as turbine blades.

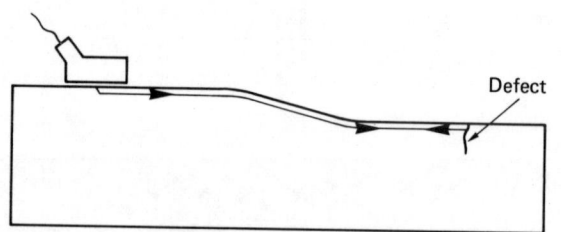

FIGURE 5.10   Crack detection using a surface wave probe.

5.13 Identification of defects

By means of ultrasonic methods not only can the exact position of internal defects be determined but it is also possible, in many cases, to distinguish the type of defect. In this section the various types of signal response received from particular types of defect will be considered.

(a) *Defect at right angles to the beam direction*

When no defect is present, a large echo signal should be received from the back-wall. The presence of a small defect should give a small defect echo and some reduction in the strength of the backwall echo. When the defect size is greater than the probe diameter the defect echo will be large and the backwall echo may be lost (figure 5.11), depending on the depth of the defect in relation to beam spread in the far zone.

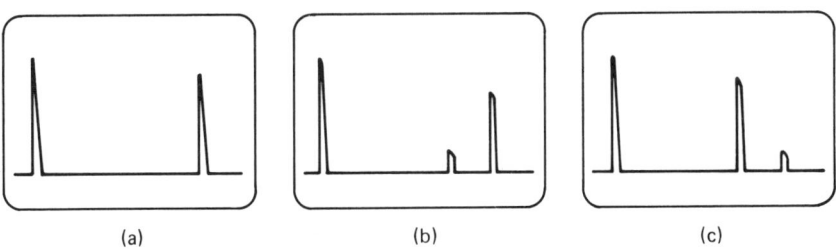

(a) (b) (c)

FIGURE 5.11   Effect of defect size on screen display: (a) defect free — initial pulse and backwall echo only; (b) small defect echo but large backwall echo; (c) large defect echo with small backwall echo.

(b) *Defects other than plane defects*

Areas of micro-porosity will cause a general scattering of the beam, giving some 'grass' on the CRT trace and with loss of the backwall echo (figure 5.12a). A large spherical or elliptical inclusion or hole would tend to give a small defect echo coupled with a small backwall echo (figure 5.12b), while a plain trace showing no echo at all could be an indication of a plane defect at some angle other than normal to the path of the beam (figure 5.12c).

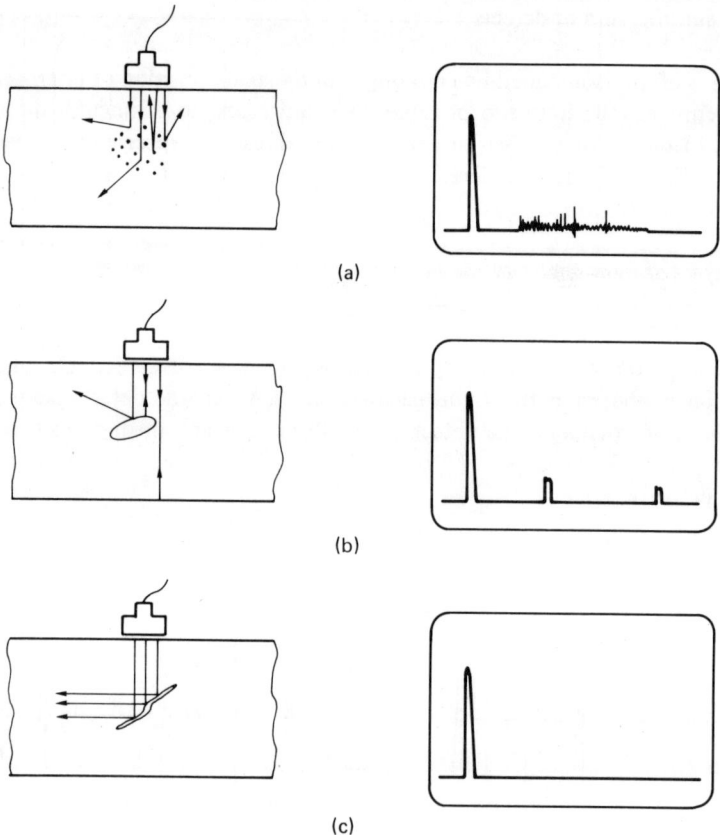

(a)

(b)

(c)

FIGURE 5.12   (a) Micro-porosity. (b) Elliptical defect. (c) Angled defect.

(c) *Laminations in thick plate*

The plate should be completely scanned in a methodical manner, as shown in figure 5.13. The indications of laminations are a closer spacing of echoes and a more rapid fall-off in the size of the echo signals. Either or both of these indications will give indication of lamination (see figure 5.14).

(d) *Lamination in thin plate*

A thin plate may be considered to be a plate of thickness less than the dead zone of the probe. A sound plate will show a regular series of echoes with exponential fall-off of amplitude. A laminated region will show a close spacing with a much faster rate of amplitude fall-off. The pattern may change from an even to an irregular outline. It is this pattern change which, in many cases, gives the best indication of lamination in thin plate (figure 5.15).

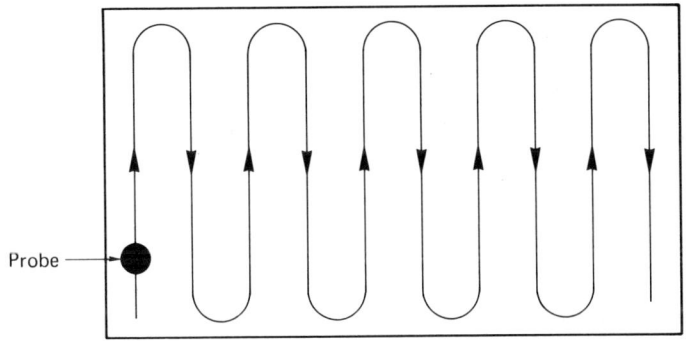

FIGURE 5.13   Method of scanning a large surface.

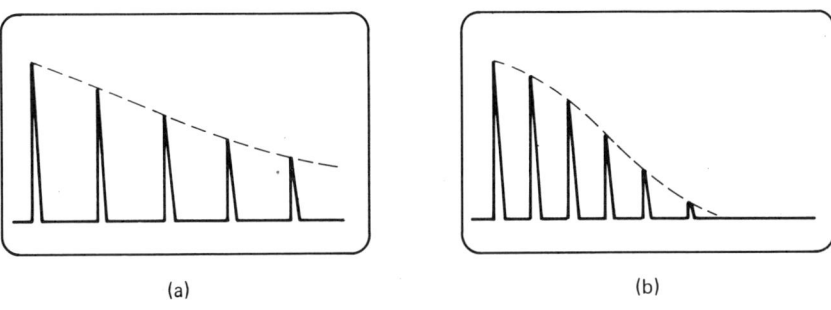

FIGURE 5.14   Indication of lamination in thick plate: (a) good plate; (b) laminated plate.

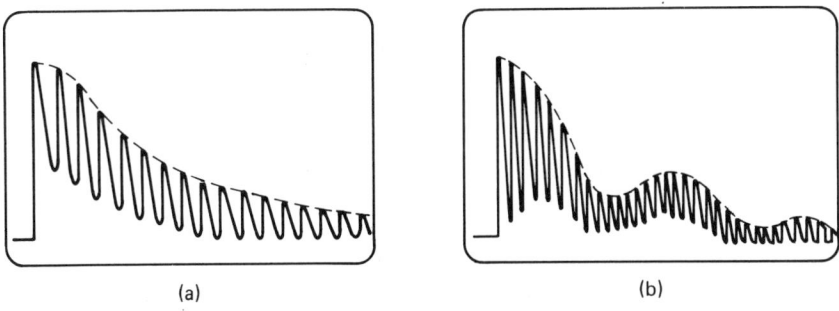

FIGURE 5.15   Indication of lamination in thin plate: (a) good plate; (b) laminated plate.

(e) *Weld defects*

Ultrasonic testing using angle probes in either the reflection or transmission mode is a reliable method for the detection of defects in butt welds and for determining their exact location. It is, however, fairly difficult to determine with certainty the exact nature of the defect, and much depends upon the skill and experience of the operator. If, following ultrasonic inspection, there is any doubt in the mind of the operator about the quality of a weld, then it would be wise to check radiographically the suspect area.

(f) *Radial defects in cylindrical tubes and shafts*

A radial defect in a cylindrical member is not generally detectable using normal probe inspection, as the defect will be parallel to the ultrasonic beam. In these circumstances the use of an angle probe reflection technique will clearly show the presence of defects (figure 5.16).

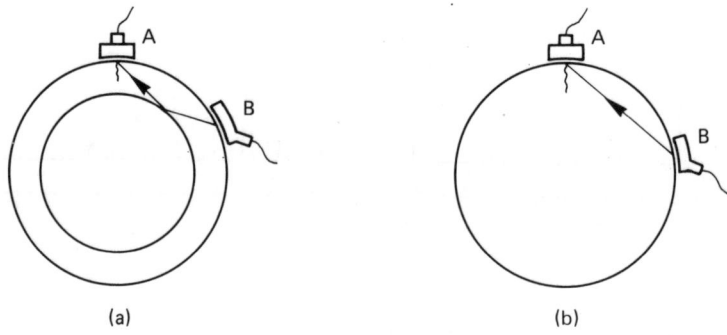

(a)   (b)

FIGURE 5.16   Detection of radial defects in: (a) tubes, (b) solid bar — normal probe in position A will not detect defect but angle probe at B will.

(g) *Bonding defects*

Ultrasonic inspection is suitable for checking the quality of bonding between either two metal surfaces or the interface between a metal and a non-metal. For the inspection of metal-to-metal bonds, scanning should always be carried out from the surface of the thinner of the two metals. If a good bond exists the ultrasonic energy will pass from one metal to the other with very little reflection of the signal, but if the bonding is poor there will be considerable reflection at the interface and a series of multiple interface echoes will be seen on the screen display (see figure 5.17).

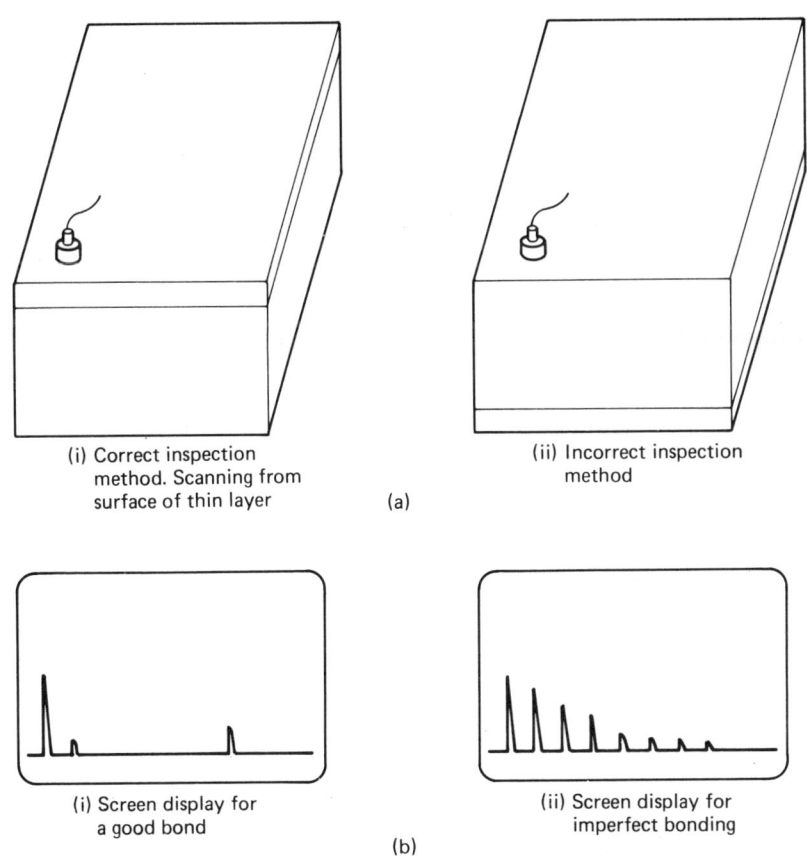

(i) Correct inspection
method. Scanning from
surface of thin layer

(ii) Incorrect inspection
method

(a)

(i) Screen display for
a good bond

(ii) Screen display for
imperfect bonding

(b)

FIGURE 5.17 (a) Inspection technique for metal-to-metal bonding. (b) Screen
representation for bonded metals.

When checking the quality of the bonding between a metal and a non-metal,
the scanning probe should always be in contact with the metal surface. Ultrasonic
energy is dissipated more rapidly within a non-metallic material than within a
metal and so there will be fairly strong absorption of signal energy at each succes-
sive reflection from the bond interface. If, however, there is poor bonding, then
there will be almost perfect reflection of signal from the interface and less attenuation of multiple interface echoes (see figure 5.18).

For the inspection of a composite structure where a non-metallic material is
sandwiched between two metal layers, it is necessary to scan from each metal
surface.

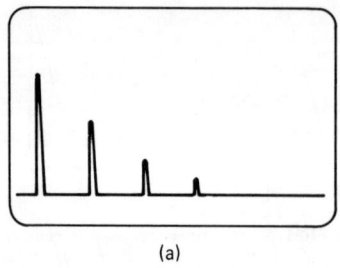 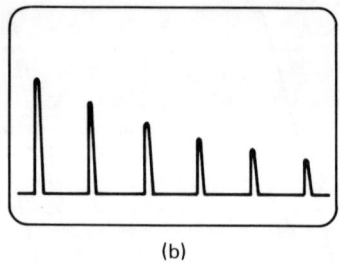

(a)   (b)

FIGURE 5.18   Metal to non-metal bonding — screen display: (a) good bond —
              rapid attenuation of multiple echoes results from strong absorp-
              tion of energy at bond interface; (b) poor bond — strong reflection
              at metal/non-metal interface giving little attenuation of signal.

While all information given in the sections (a) to (g) above gives a guide to the
types of 'A' scan response which can be obtained in various situations, it should
be emphasised that an operator needs to have a sound training and considerable
experience before he will be able to give a skilled interpretation of 'A' scan images.

5.14 Immersion testing

All the techniques of inspection which have been discussed so far are of the type
known as 'contact scanning' in which the inspection probe or probes are held in
contact with the surface of the material, through a thin film of liquid couplant. It
is also possible to use the system of 'immersion scanning' in which the component
to be inspected is immersed in a tank of water and the test probe is placed above
the testpiece but below the water surface. Immersion testing is ideally suited for
the examination of part processed or finished parts in a production plant and the
test equipment is usually fully automated. Examples of its use are for the inspection of slabs before these are machined into aircraft structural components, for
the inspection of gas turbine discs, and for the inspection of aircraft wheels. In the
case of the last mentioned item, a large airline may use immersion testing for the
routine maintenance inspections.

   The screen display obtained during an immersion test will show a blip corresponding to the water/metal interface, a backwall echo from the material and,
between these two, a blip corresponding to any defect which may be present
(figure 5.19). The distance between the probe and test material must be set so
that repeat echoes from the water/metal interface do not appear within the length
of the time base corresponding to the thickness of metal. A time-base delay is
usually incorporated into the display so that the first peak visible at the left-hand

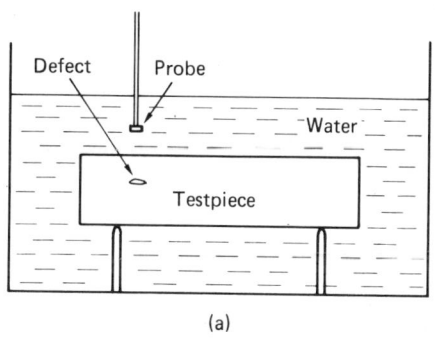

(a)

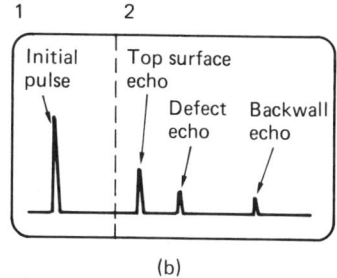

(b)

FIGURE 5.19   Immersion testing: (a) testpiece arrangement; (b) screen display —
with time base delay, start of trace can be transposed from 1 to 2.

edge of the screen corresponds to the metal/water interface. Most of the automated equipment designed for immersion testing also produces a permanent chart
record of the testpiece showing the position of defects using either a 'B' scan or
'C' scan type display.

5.15 Sensitivity and calibration

The sensitivity of ultrasonic testing is related to the frequency of the waves, as
stated in section 5.4, but it is also affected by adjustments to the oscilloscope controls. If the sensitivity is too high, natural features of the material, such as the
large grain size of castings, will cause echoes and these may mask the presence of
defects. The frequency to be used must be chosen carefully. High frequencies are
suitable for fine-grained wrought materials.

In many cases the fact that a defect is present does not necessarily mean that a
component must be thrown out. Defects may be tolerated provided that they do
not exceed a certain acceptable size. One of the benefits of ultrasonic inspection is

that defect size can be accurately determined. This means that a component containing, say, a small crack can remain in service and further inspection at set intervals can be used to monitor the growth of the crack until its size is no longer acceptable and the part can be withdrawn. In order to be able to determine defect size accurately, the oscilloscope must be calibrated and this is done with the aid of testpieces containing artificial defects of certain specific sizes.

5.16 Reference standards

As stated earlier in sections 5.7 and 5.8, there are numerous factors which affect the propagation of an ultrasonic wave pulse during component inspections, and hence, exact quantitative determination of defect size, shape and orientation can be difficult. Test probes and ultrasonic equipment can be calibrated by using reference blocks and calibration standards. This will permit the evaluation and recording of quantitative data on defects.

When selecting standard reference blocks there are several variables to be considered. These include the nature of the component material and the effects of thermal and mechanical processing on the acoustic properties of the component material, the component shape and, hence, the direction of the ultrasonic beam, depth, amplitude variations of the beam and the nature of the flaw to be detected. Three types of standard reference blocks are normally employed for calibration of the test equipment: distance-amplitude blocks, area–amplitude blocks and test blocks recommended by the International Institute of Welding (IIW–type reference blocks).

Distance-amplitude blocks

Distance-amplitude blocks are used to determine the relationship between depth and amplitude for direct beam inspection in a given material. Such standards are made from cylindrical bars of the same material as that of the testpiece. Artificial flat-bottomed holes of a specific size and of varying depth are drilled into the reference standard from the base. Because the decrease in echo amplitude from a flat-bottomed hole using a circular search unit is inversely proportional to the square of the distance to the bottom of the hole, it is possible to apply a distance-amplitude correction to the test equipment, so that a flaw of a given size will produce an indication on an oscilloscope screen that is of a pre-determined height regardless of the distance from the entry surface. Normally, a number of reference standards of this type are made, with artificial 'flaws' of varying size.

Area–amplitude blocks

The amplitude of an echo from a flat-bottomed hole in the far field of a direct beam search probe is proportional to the area of the bottom of the hole. Area-amplitude blocks are made from material with the same acoustic properties as that of the test component. Artificial flaws of different sizes are machined to the same depth in the reference block. These standards can be used to relate echo amplitude to the size of a flaw, and to check the linearity of a pulse-echo inspection system with respect to flaw size. However, because natural flaws do not correspond directly to the regular shape of a flat bottomed hole and, hence, are less than ideal in reflective properties, an area–amplitude block defines only a lower limit for the size of a flaw which yields a given amplitude of indication. Often, it is not possible to determine accurately the true size of real flaws.

Calibration blocks

IIW steel calibration blocks are used universally for the calibration of both angle beam and direct beam probes prior to contact inspection of components. A typical block design is shown in figure 5.20.

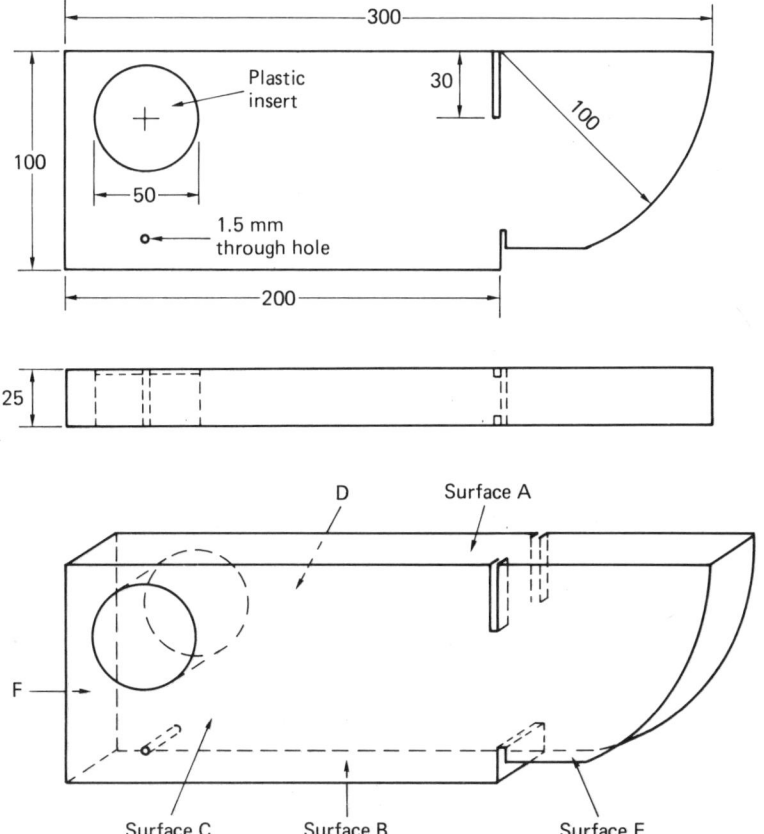

FIGURE 5.20   Type A2 IIW calibration block (dimensions in mm)

Index point and beam spread

To determine the index point of an angle beam probe, that is the point where the beam leaves the probe, the probe is placed on surface A of the calibration block in the position shown in figure 5.21.

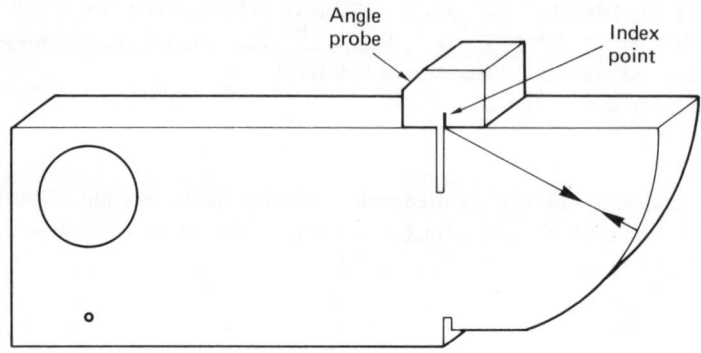

FIGURE 5.21   Determination of the index point of an angle-beam probe using the IIW type A2 test-block.

The probe is moved along the block surface until an echo of maximum amplitude is attained. Regardless of the 'angle' of a probe, a maximum echo can only be obtained when the index point of the probe coincides with the 'focal point' of the curved surface of the block. Hence, the point on the probe that is directly over the focal point of the block can be marked as the index point.

Once the index point of an angle probe has been marked, beam spread can be determined. The probe is moved in one direction away from the focal point of the block along surface A, in either direction, until the echo disappears. The beam angle at the index point is noted. The probe is then moved in the opposite direction, past the focal point of the block, until the echo disappears again. The beam spread is the difference between the angles indicated by the index point at these extreme positions.

Beam angle

In order to determine the 'beam angle' of an angle probe, the probe is placed on surface A or surface B and is 'aimed' toward the 50 mm diameter hole, as shown in figure 5.22.

The probe is moved along the block surface until a maximum amplitude echo is received. At this position, the index point indicates the 'beam angle' which can be read from the scales marked along the sides of the block at the edges of surface A and surface **B**. The beam angle so determined is the angle of the refracted shear wave, produced by a given angle probe, in the low-carbon steel calibration block.

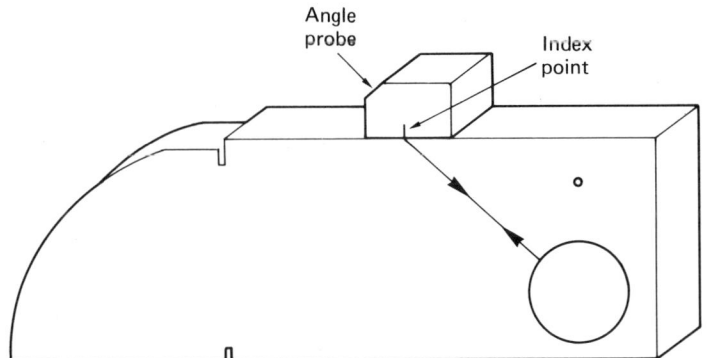

FIGURE 5.22   Beam angle determination using IIW type A2 test-block.

Time base calibrations

In order to obtain an accurate measurement of the position of a defect within a component by ultrasonic pulse-echo inspection, it is necessary to calibrate the oscilloscope time base prior to inspection.

For a direct beam probe, the time base calibration can be achieved by placing the probe on either surface **C** or surface **D** of the calibration block, to obtain multiple echoes from the 25 mm thickness. The time base can be adjusted until the echoes corresponding to 50, 100, 150, 200, 250 mm, and so on, are aligned with appropriate grid lines on the oscilloscope screen. Upon completion of these adjustments the instrument time base is calibrated in terms of 'metal distance' in steel.

The calibration for steel can be converted to a calibration for another material by multiplying 'metal distance' in steel by the ratio of the velocity of sound in the material to the velocity of sound in steel.

The time base calibration procedure for angle probes can be achieved using a similar procedure to that employed for direct beam probes. The angle beam probe is placed on surface **A** with the index point of the probe and focal point of the block coinciding. Multiple reflections are obtained from the 100 mm radius. The oscilloscope screen time base can be adjusted as described previously, to obtain the desired calibration.

Calibration of the time base, for either type of probe, allows an estimate of the probe 'dead zone' to be obtained. The dead zone is estimated by measuring the width of the initial pulse indication at its base, as shown schematically in figure 5.23.

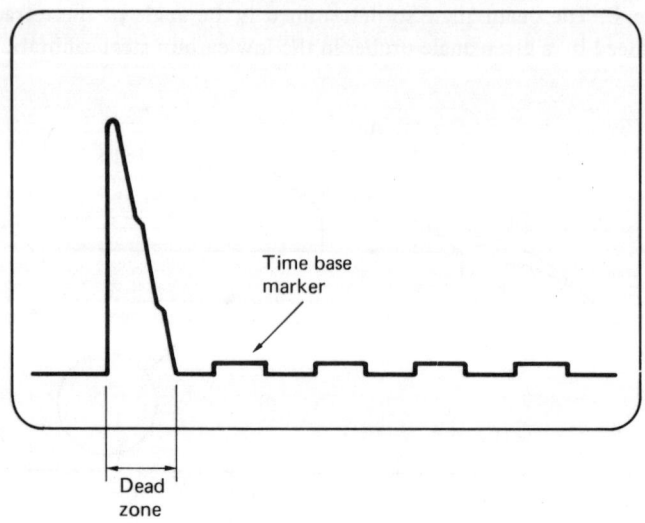

FIGURE 5.23   Dead zone estimation from time base calibration.

Screen resolution

The ability of a recording instrument to resolve and separate back reflection echoes is affected by the characteristics of both the instrument amplifier and the ultrasonic probe employed. Hence, it is advisable to check this resolution, prior to component inspection, for any combination of probe and instrument.

For direct beam probes, the standard IIW Type A2 calibration block provides a facility for implementing such checks. The probe is placed on surface A in the position shown in figure 5.24. If the resolution is sufficiently good, there should

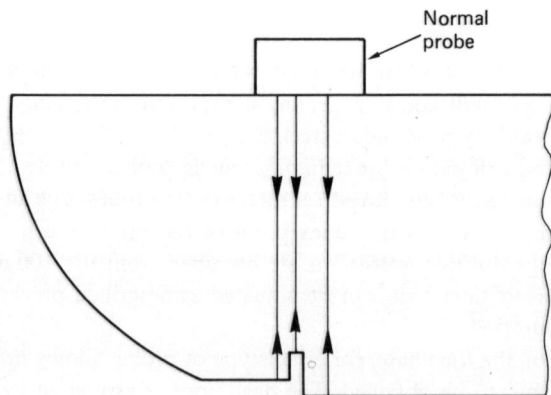

FIGURE 5.24   Determination of back-surface resolution for a direct beam probe.

be echo indications from the base of the 2 mm wide notch and from the two block surfaces, B and E, on either side of the notch. For good resolution, these three echoes should be clearly separated and not overlapped.

Sensitivity

The relative sensitivity of a given probe and instrument combination can be determined by using the 1.5 mm through hole in the type A2 calibration block (see figure 5.20).

For a direct beam probe, the sensitivity can be defined by placing the probe on either surface B or surface F in line with the hole. The position of the probe is adjusted until a maximum echo indication is received. The instrument gain control may then be adjusted to give the desired indication height on the oscilloscope screen, For an angle probe a similar procedure is followed, but with the probe positioned on either surface C or surface D.

Thickness blocks

Ultrasonic test equipment may be calibrated for thickness measurements using step wedges or tapered test blocks. Ideally, reference blocks of this type should be manufactured from material similar to that of the component to be inspected, and the exact thickness of the various test block sections should be determined with mechanical gauges and marked on the block. Alternatively, commercial reference standards of this type, made of normalised bars of carbon steel, are available. Obviously, the difference in acoustic properties between the material of the test component and the steel of the test block must be determined experimentally, in order to attain accurate thickness measurements.

5.17 Surface condition

Ideally, a smooth surface is required on a material or component for effective ultrasonic inspection. The rough surfaces of castings or forgings may present a problem. The use of a thick grease as a couplant may overcome this, or the parts can be tested using the water immersion method. Rust, surface scale and loose paint should always be removed from a metal surface before testing, otherwise poor results may be obtained.

If it is desired to test components with a contoured surface, it will probably be necessary to specially manufacture a test probe, with the transducer crystal embedded in a perspex slipper which is shaped to match the contours of the components.

5.18 Some applications of ultrasonic testing

As has been seen in the foregoing paragraphs, ultrasonic test methods are suitable for the detection, identification and size assessment of a wide variety of both surface and sub-surface defects in metallic materials, provided that there is, for reflection techniques, access to one surface. There are automated systems which are highly suitable for the routine inspection of production items at both an intermediate stage and the final stage of manufacture. Using hand-held probes, many types of component can be tested, including *in situ* testing. This latter capability makes the method particularly attractive for the routine inspection of aircraft and road and rail vehicles in the search for incipient fatigue cracks. In aircraft inspection, specific test methods have been developed for each particular application and the procedures listed in the appropriate manuals must be followed if consistent results are to be achieved. In many cases a probe will be specially designed for one specific type of inspection.

Present-day ultrasonic equipment is compact and light and will operate from either a standard mains supply or from its internal battery. A typical set would have dimensions of about 300 mm × 250 mm × 100 mm and weigh less than 5 kg. The equipment is extremely portable, relatively inexpensive and extremely versatile, and this has helped ultrasonic testing to become an indispensable tool for those concerned with all aspects of quality control and quality assurance.

Figure 5.25 shows the inspection of large forged bars using a compression wave technique, and figure 5.26 shows the use of an angle probe for the checking of circumferential welds in a large diameter pipeline. Both illustrations show the compact size and portable nature of ultrasonic test equipment.

FIGURE 5.25   Ultrasonic inspection of forged bars using a compression wave probe (courtesy of Wells–Krautkramer Ltd).

FIGURE 5.26   Ultrasonic inspection of pipeline weld using a shear wave probe
(courtesy of Wells–Krautkramer Ltd).

Figure 5.27 also shows an angled shear wave probe in position on a section of
plate containing a butt weld.

As mentioned earlier, accurate thickness measurements can be made using an
ultrasonic pulse-echo technique. Some ultrasonic equipment is made specifically
for thickness measurements and instead of embodying a cathode ray tube for
signal display they give a direct digital read-out of section thickness in millimetres
or in inches. Figure 5.28 shows an instrument of this type in use to check the wall
thickness of a valve body in a chemical plant as part of a routine maintenance
inspection. The sensitivity of this type of instrument is such that it can give thickness readings over the range from 1.25 mm to 1000 mm with an accuracy of ±0.02
per cent. This degree of accuracy permits the monitoring of corrosion or erosion
by noting small changes in wall thickness.

While it was stated in section 5.7 that sound cannot be transmitted easily across
an air gap and that a fluid couplant is required between the transducer and the
testpiece, it is possible, in certain cases, for a special type of transducer to function

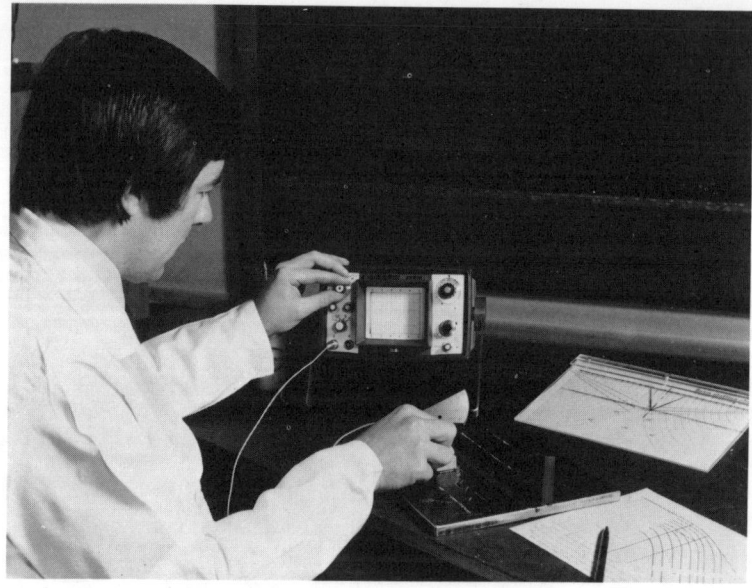

FIGURE 5.27   Ultrasonic flaw detector being set up in a laboratory for weld
               inspection (courtesy of Wells–Krautkramer Ltd).

FIGURE 5.28   Maintenance check of valve wall thickness in a chemical plant
               using a digital read-out ultrasonic wall thickness meter (courtesy
               of Inspection Instruments (NDT) Ltd and Sonic Instruments
               Inc. (USA)).

without a couplant. Figure 5.29 shows a special-purpose test rig for measuring wall thickness and determining any eccentricity in hot-forged steel tube. Specially designed water-cooled transducer heads which can withstand steel surface temperatures in excess of 800°C and which can function without a couplant are used. The tube wall thickness is measured at four points at 90° intervals around the circumference. The results of each measurement are given on an automatic printer as well as being displayed on four separate cathode ray tubes.

Soft-face probes may also be used without a couplant. This type of probe is faced with a soft polyurethane which will deform and adjust to conform to a rough surface. One of the applications for which this type of probe is used is checking the quality of welds in the top bars of cells for electrical storage batteries.

Electro-magnetic acoustic transducers offer another means of avoiding the use of a couplant. In this case there is no direct contact between the transducer and the test component. The transducer is composed of a powerful electro-magnet and a radio-frequency coil. Interaction between the magnetic field and the induced high-frequency eddy currents generates a force which stimulates an elastic wave within the material. The separation distance between the transducer and the workpiece is of the order of a few millimetres. One of the applications of this type of ultrasonic generator is for the measurement of wall thinning in corroded boiler tubes.

FIGURE 5.29   Four-channel ultrasonic installation for wall thickness and eccentricity measurements on hot-forged steel tube. Measurements can be made when the tube surface temperature is as high as 815°C (courtesy of Inspection Instruments (NDT) Ltd and Sonic Instruments Inc. (USA)).

6

Radiography

6.1 Introduction

Very-short-wavelength electromagnetic radiation, namely X-rays or γ-rays, will penetrate through solid media but will be partially absorbed by the medium. The amount of absorption which will occur will depend upon the density and thickness of the material the radiation is passing through, and also the characteristics of the radiation. The radiation which passes through the material can be detected and recorded on either film or sensitised paper, viewed on a fluorescent screen, or detected and monitored by electronic sensing equipment. Strictly speaking, the term *radiography* implies a process in which an image is produced on film. When a permanent image is produced on radiation-sensitive paper, the process is known as *paper radiography*. The system in which a latent image is created on an electro-statically charged plate and this latent image used to produce a permanent image on paper is known as *xeroradiography*. The process in which a transient image is produced on a fluorescent screen is termed *fluoroscopy*, and when the intensity of the radiation passing through a material is monitored by electronic equipment the process is termed *radiation gauging*.

It is possible to ultilise a beam of neutrons rather than X-rays or γ-rays for inspection purposes, this being termed *neutron radiography* (refer to section 7.2).

After an exposed radiographic film has been developed, an image of varying density will be observed with those portions of the film which have received the largest amounts of radiation being the darkest. As mentioned earlier, the amount of radiation absorbed by the material will be a function of its density and thickness. The amount of absorption will also be affected by the presence of certain defects such as voids or porosity within the material. Thus radiography can be used for the inspection of materials and components to detect certain types of defect.

The use of radiography and related processes must be strictly controlled because exposure of humans to radiation could lead to body tissue damage.

6.2 Uses of radiography

Radiography is capable of detecting any feature in a component or structure provided that there are sufficient differences in thickness or density within the testpiece. Large differences are more readily detected than small differences. The main types of defect which can be distinguished are porosity and other voids and inclusions, where the density of the inclusion differs from that of the basis material. Generally speaking, the best results will be obtained when the defect has an appreciable thickness in a direction parallel to the radiation beam. Plane defects such as cracks are not always detectable and the ability to locate a crack will depend upon its orientation to the beam. The sensitivity possible in radiography depends upon many factors but, generally, if a feature causes a change in absorption of 2 per cent or more compared with the surrounding material, then it will be detectable.

Radiography and ultrasonics (see chapter 5) are the two methods which are generally used for the successful detection of internal flaws that are located well below the surface, but neither method is restricted to the detection of this type of defect. The methods are complementary to one another in that radiography tends to be more effective when flaws are non-planar in type, whereas ultrasonics tends to be more effective when the defects are planar.

Radiographic inspection techniques are frequently used for the checking of welds and castings, and in many instances radiography is specified for the inspection of components. This is the case for weldments and thick-wall castings which form part of high-pressure systems.

Radiography can also be used to inspect assemblies to check the condition and proper placement of components. It is also used to check the level of liquid in sealed liquid-filled systems. One application for which radiography is very well suited is the inspection of electrical and electronic component assemblies to detect cracks, broken wires, missing or misplaced components and unsoldered connections.

Radiography can be used to inspect most types of solid material but there could be problems with very high or very low density materials. Non-metallic and metallic materials, both ferrous and non-ferrous, can be radiographed and there is a fairly wide range of material thicknesses that can be inspected. The sensitivities of the radiography processes are affected by a number of factors, including the type and geometry of the material and the type of flaw, and these factors will be discussed in later sections.

6.3 Some limitations of radiography

Although radiography is a very useful non-destructive test system, it possesses some relatively unattractive features. It tends to be an expensive technique, compared with other non-destructive test methods. The capital costs of fixed X-ray equipment are high but coupled with this considerable space is needed for a radiography laboratory, including a dark room for film processing. Capital costs will be much less if portable X-ray sets or γ-ray sources are used for *in situ* inspections, but space will still be required for film processing and interpretation.

The operating costs for radiography are also high. The setting-up time for radiography is often lengthy and may account for over half of the total inspection time. Radiographic inspection of components or structures out on sites may be a lengthy process because the portable X-ray equipment is usually limited to a relatively low energy radiation emission. Similarly, portable radio-active sources emitting γ-radiation tend to be of fairly low intensity. This is because high-intensity sources require very heavy shielding and thus cease to be truly portable. In consequence, on-site radiography in the field tends to be restricted to a maximum material thickness of 75 mm of steel, or its equivalent. Even then, exposure times of several hours may be needed for the examination of thick sections. This brings a further disadvantage in that personnel may have to be away from their normal work posts for a long time while radiography is taking place.

The operating costs for X-ray fluoroscopy are generally much lower than those for radiography. Setting-up times are much shorter, exposure times are usually short and there is no need for a film processing laboratory.

Another aspect which adds to radiography costs is the need to protect personnel from the effects of radiation, and stringent safety precautions have to be employed. This safety aspect will apply to all who work in the vicinity of a radiography test as well as to those persons directly concerned in the testing.

As mentioned in section 6.2, it is not possible to detect all types of defect by means of radiography. Cracks may only be detected if they lie in a direction parallel to the X-ray beam and even then tight cracks may escape detection. Lamination type defects in metals are almost impossible to detect by means of radiography.

6.4 Principles of radiography

The basic principle of radiographic inspection is that the object to be examined is placed in the path of a beam of radiation from an X-ray or γ-ray source. A recording medium, usually film, is placed close to the object being examined but on the opposite side from the beam source (as shown in figure 6.1). X or gamma radiation cannot be focused as visible light can be focused and, in many instances the radiation will come from the source as a conical beam. Some of the radiation will be absorbed by the object but some will travel through the object and impinge on the film, producing a latent image. If the object contains a flaw which has different absorptive power from that of the object material, the amount of radiation emerging from the object directly beneath the flaw will differ from that emerging from adjacent flaw-free regions. When the film has been developed there will be an area of different image density which corresponds to the flaw in the material. Thus the

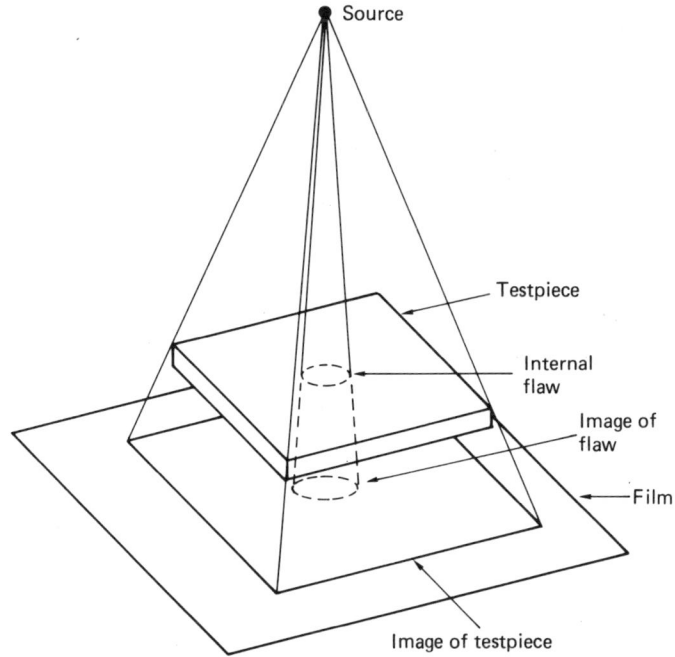

FIGURE 6.1   Schematic diagram showing radiographic system.

flaw will be seen as a shadow within the developed radiograph. This shadow may be of lesser or greater density than the surrounding image, depending on the nature of the defect and its relative absorptive characteristics.

The developed radiograph is a two-dimensional representation of a three-dimensional object and the image may be distorted both in size and shape compared with the testpiece. The reasons for this distortion are given in section 6.11.

The position of a flaw within a testpiece cannot be determined exactly with a single radiograph but by taking several radiographs with the beam directed at the object from a different angle for each exposure, it should be possible to determine the exact position of the flaw in relation to the thickness of the object.

6.5 Radiation sources

A portion of the electro-magnetic spectrum is shown in figure 6.2.

The very-high-frequency (short wavelength) radiation, listed as X-rays and γ-rays, is the only form of electro-magnetic radiation which will penetrate solid and opaque material. Electro-magnetic waves can be regarded as a series of quanta

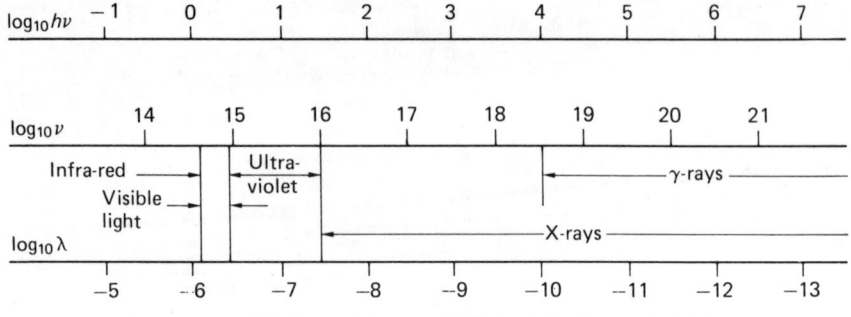

FIGURE 6.2   Portion of the electro-magnetic spectrum.

or photons, and the energy of the photons varies with the frequency of the radiation. The relationship between frequency and photon energy is given by Planck's
quantum relationship $E = h\nu$, where E is photon energy, ν is frequency and h is
Planck's constant ($h = 6.625 \times 10^{-34}$ J s). The energies of photons of radiation in
the X-ray and γ-ray sector of the spectrum range from 50 to 10^6 or 10^7 electron
volts. (The electron volt (eV) is a unit of energy and is the energy required to
move one electron through a potential difference of one volt. 1 eV = 1.602 $\times$
10^{-19} J.) The photon energies for radiation of different frequencies is shown in
figure 6.2.

X-rays and γ-rays are indistinguishable from one another. The only difference
between them is the manner of their formation. X-rays are formed by bombarding
a metal target material with a stream of high-volocity electrons within an X-ray
tube. γ-rays, on the other hand, are emitted as part of the decay process of radio-
active substances.

6.6 Production of X-rays

As mentioned earlier, X-rays are produced by bombarding a metal target with a
beam of high-velocity electrons. The major components of an X-ray tube are a
cathode to emit electrons and an anode target, both being contained within an
evacuated tube or envelope. The general arrangement is shown in figure 6.3. The
cathode is a filament coil of tungsten wire. An electric current at a low voltage
flows through the cathode filament to heat it to incandescence and stimulate the
thermionic emission of electrons. A large electrical potential difference (the tube
voltage) exists between the cathode and anode target to accelerate the electrons
across the space separating the two. X-ray tube voltages generally range from
50 kV to 1 MV.

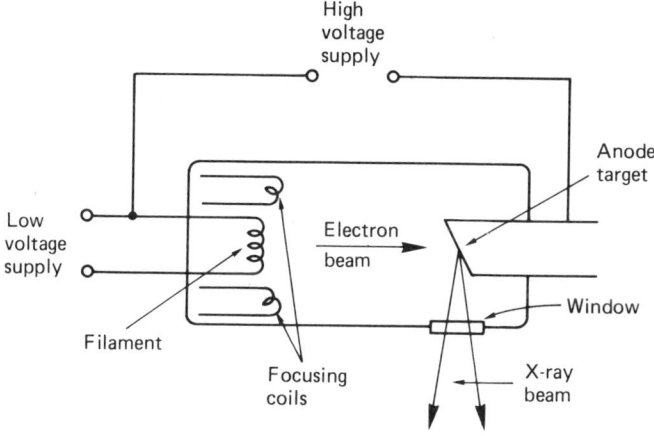

FIGURE 6.3   Schematic view of an X-ray tube.

A focusing cup or focusing coil is placed close to the cathode. This acts as an electro-magnetic lens to focus the thermionic emission into a fine beam aimed at the centre of the anode target material.

The anode comprises a small piece of the target metal, which is usually tungsten, embedded in a mass of copper. Tungsten is used as a target material because it is a highly efficient emitter of X-rays and because it possesses an extremely high melting point, 3380°C, and can therefore withstand the high temperatures generated by the impinging electrons. The tungsten target material is embedded in a copper block which is water-cooled or oil-cooled so that the heat energy generated can be readily dissipated by conduction through the copper.

The X-ray tube envelope may be made of glass, a ceramic material such as alumina, a metal, or a combination of materials. The majority of X-ray tubes made these days are of ceramic/metal construction and these can be made smaller, for any particular tube voltage rating, than glass/metal tubes. The tube envelope must possess good structural strength at high temperatures in order to withstand the combined effects of radiated heat from the anode and the forces exerted by atmospheric pressure on the evacuated chamber. The shape of the envelope may vary with the tube voltage rating and the nature of the anode and cathode design. The envelope must contain a window opposite the anode to permit the X-ray beam to leave the tube. The window is made of a low atomic number element to minimise X-ray absorption. A 3 to 4 mm thickness of beryllium is generally used as a window material. The electrical connections for anode and cathode are fused into the walls of the envelope.

The X-ray tube is contained within a metal housing which is well insulated to give protection from high-voltage electrical shock, and this housing usually possesses a high-voltage plug and socket which will permit the rapid disconnection of the electrical cables which connect the tube to the high-voltage generator unit.

The portable X-ray units which are used for on-site work are generally self-contained with both the high-voltage generator and the X-ray tube contained within the same housing. In this case there are no high-voltage cables outside the unit.

A low-voltage current passes through the cathode filament, heating it and generating a cloud of electrons around the filament by thermionic emission. When a high voltage is applied across the tube between cathode and anode, the electrons are accelerated across the evacuated space to strike the target. The electron beam is focused so that it impinges on a small area of the target, this small area being termed the 'focal spot'.

Most of the electron beam energy is converted into heat energy, which has to be dissipated — some is converted into X-radiation. The smaller the focal spot on the target, the sharper will be the radiographic image that can be obtained. However, the extent of anode heating which occurs prevents the use of too small a focal spot. The anode and target design is a compromise between the conflicting requirements of long target life and maximum radiographic definition. In many X-ray tube designs, the face of the anode is inclined at an angle to the electron beam. The electron beam is focused in such a way as would give a small square focal spot on a plane normal to the electron beam but which is a long, narrow focal spot on the inclined target face (see figure 6.4).

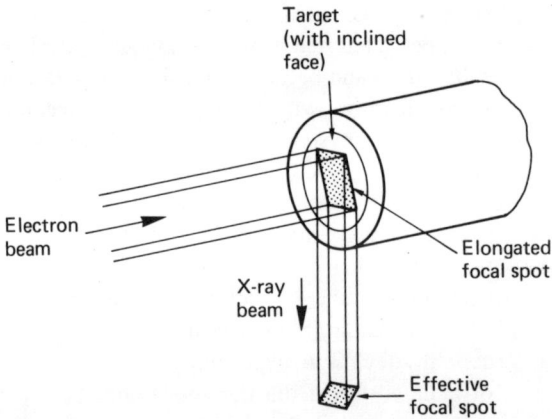

FIGURE 6.4   Diagram showing actual and effective focal spots for an anode target with an inclined face.

There are three important variables in X-ray tubes: these are the filament current, the tube voltage and the tube current. A change in the filament current will alter the temperature of the filament which will change the rate of thermionic emission of electrons. An increase in the tube voltage, the potential difference between cathode and anode, will increase the energy of the electron beam and, hence, will increase the energy and penetrating power of the X-ray beam which is produced. The third variable, the tube current, is the magnitude of the electron flow between cathode and anode and is directly related to the filament temperature. (The tube current is usually referred to as the milli-amperage of the tube.) The intensity* of the X-ray beam produced by the tube is approximately proportional to the tube milli-amperage.

Figure 6.5 shows an industrial X-ray set with a maximum rated tube voltage of 320 kV. This type of set is suitable for use in an X-ray test laboratory. The metal/ceramic X-ray tube, within its shielding, is mounted on an adjustable stand which

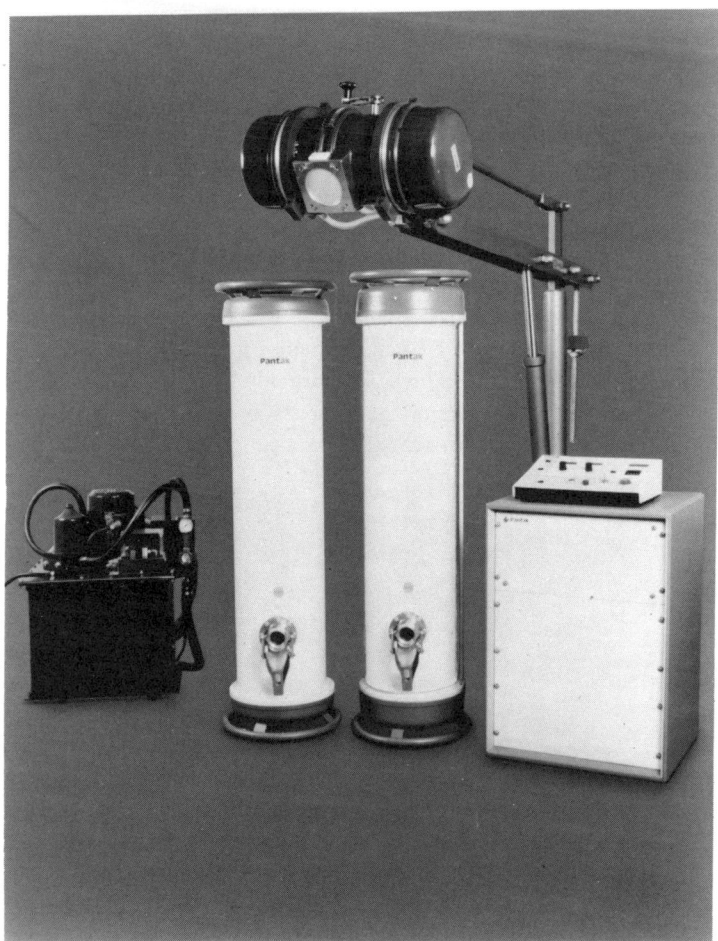

FIGURE 6.5   320 kV X-ray unit (courtesy of Pantak Ltd).

---

* The intensity of X-radiation or γ-radiation is measured in roentgens. One roentgen is the amount of radiation that will produce ions carrying one electrostatic unit of electricity in 1.293 mg of air.

allows for accurate positioning of the tube in relation to the workpiece. Each of the two cylinders is a 160 kV high-voltage generator. The electronic control unit will give control of tube voltage within the range from 5 to 320 kV and tube currents ranging from 0.5 to 30 mA. On the left of the picture can be seen the oil pump and oil cooler.

6.7 X-ray spectra

The radiation generated by a particular X-ray tube is not of one single wavelength but covers a range of wavelengths. The X-rays are produced by two types of process. The extreme deceleration of electrons when they collide with atoms of the target material produces X-rays of many wavelengths. This 'white' radiation, or continuous band of frequencies in the X-ray spectrum, is referred to as *bremsstrahlung*. In addition to the production of bremsstrahlung, the collision of an electron with a target atom may displace an orbital electron within the target atom and move it into a higher energy state. This is termed *excitation*. When the high-energy orbital electron falls back to its original orbital position, the energy release will be emitted as radiation at a specific frequency. This radiation of characteristic frequency will be of a much higher intensity than the background white radiation (see figure 6.6). There will normally be more than one characteristic X-ray wavelength for any given target material but the dominant one is termed K_α radiation. The K_α wavelength produced by a tungsten target is 2.1 nm.

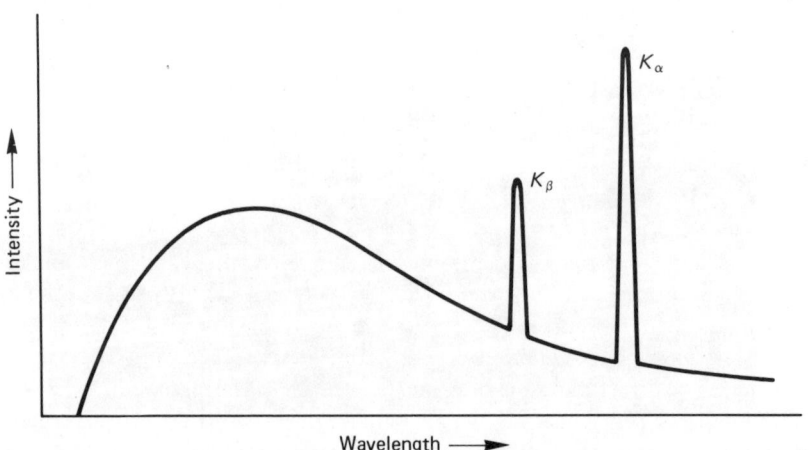

FIGURE 6.6 Spectrum of X-radiation showing characteristic frequencies and 'white' radiation.

There is a lower limit to the wavelength of the radiation produced by an X-ray tube, and this lower limit is inversely proportional to the tube voltage. The minimum wavelength, in nm, is given by

$$\lambda_{min} = \frac{1239.5}{V}$$

where V is the tube voltage.

The effects of changes in tube voltage on both the X-ray spectrum and the intensity of the radiation is shown in figure 6.7. It is the radiation at the low-wavelength end of the spectrum that is of the greatest importance in radiography because of its greater penetrating ability.

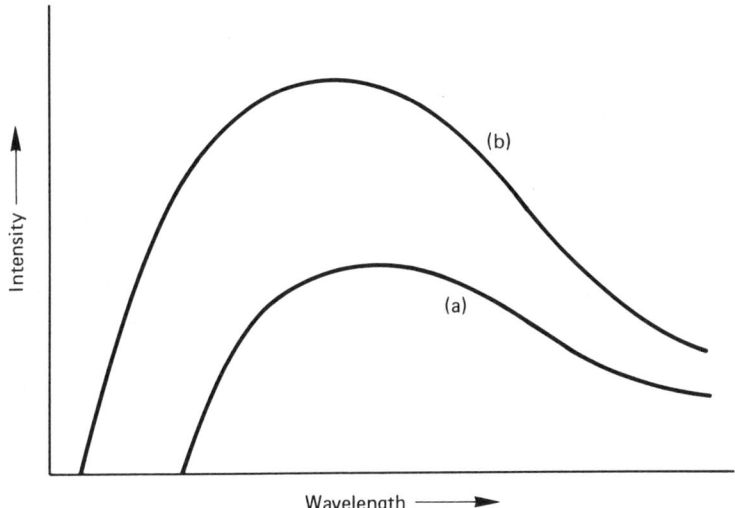

FIGURE 6.7   Effect of tube voltage on variation of radiation intensity with wavelength: (a) low tube voltage; (b) high tube voltage.

The magnitude of the tube current does not affect the wavelength spread of the X-radiation produced but it does have an effect on the intensity of the radiation. This is shown in figure 6.8.

The wavelength of X-radiation or γ-radiation is very important. The ability of the radiation to penetrate a medium increases as the wavelength reduces. In other words, compared with longer wavelength radiation, very short wavelength radiation will either penetrate a greater thickness of a given material or will be capable of penetrating materials of greater density. Hence, if the minimum wavelength of the radiation produced decreases as the tube voltage is increased, so the penetrating ability of the radiation will be increased as the tube voltage increases. This effect is shown in figure 6.9.

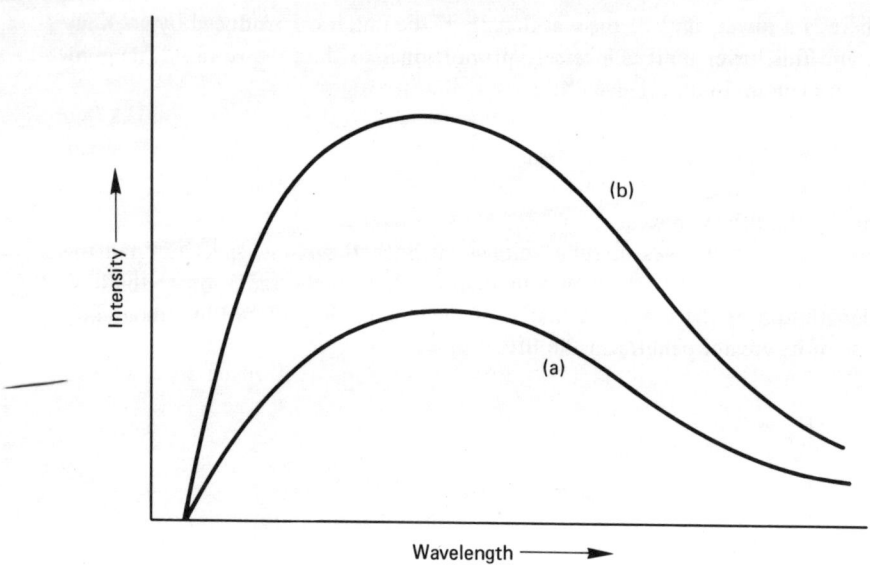

FIGURE 6.8   Effect of tube current on the variation of X-ray intensity with wavelength: (a) low tube current; (b) high tube current.

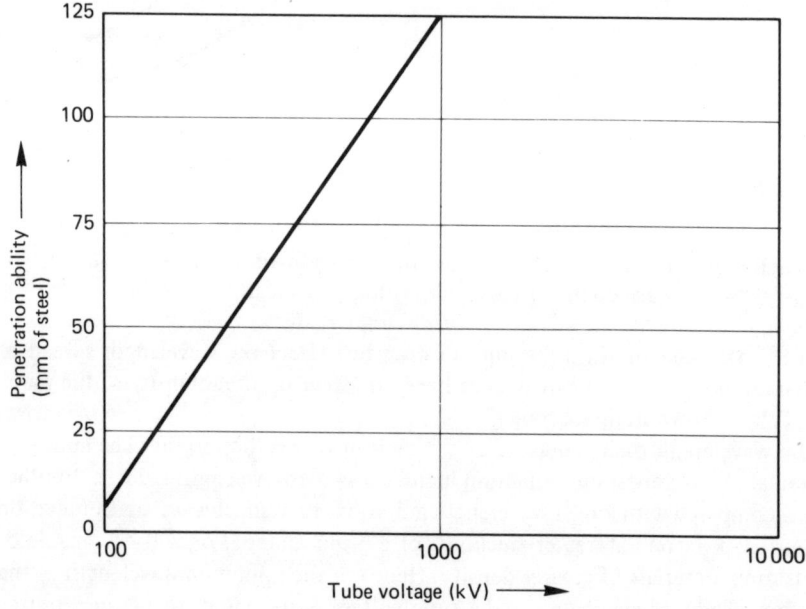

FIGURE 6.9   Effect of tube voltage on the penetrating ability of X-radiation.

It will be seen that radiation from a 200 kV tube will penetrate a steel thickness of about 25 mm. If the tube voltage is increased to 1000 kV (1 MV), the radiation will penetrate a steel thickness of about 130 mm.

The upper practical limit for conventional X-ray tubes is about 1000 kV and this will give an X-ray spectrum with the shortest wavelength radiation, possessing a photon energy of approximately 1 MeV. X-radiation with photon energies of up to 30 MeV can be produced by using high-energy electrons produced by either a van der Graaf generator, a linear accelerator or a betatron source. The penetration ability, with regard to steel, for X-rays produced in X-ray tubes and using high energy sources, is given in table 6.1. It should be noted that the maximum thickness values quoted in table 6.1 are those thicknesses which can be inspected using exposure times of several minutes' duration and medium speed film. Thicker sections can be inspected using long exposure times and a fast film speed.

Table 6.1

X-ray tubes		High-energy sources	
Tube voltage (kV)	Penetration ability (mm of steel)	Photon energy (MeV)	Penetration ability (mm of steel)
150	up to 25	2	5 to 200
250	up to 70	4.5	25 to 300
400	up to 100	7.5	60 to 450
1000	5 to 140	20	75 to 600

6.8 γ-radiation sources

γ-radiation is emitted during the decay of radio-active nuclei. Unlike the broad 'white' spectrum of radiation obtained from an X-ray tube, a γ-ray emitter will give one or more discrete radiation wavelengths, each one with its own characteristic photon energy. Radium, a naturally occurring radio-active element, has been used as a source of γ-radiation for radiography, but it is far more usual to use radio-isotopes produced in a nuclear reactor. The specific isotopes which are generally used as γ-ray sources for radiography are caesium-137, cobalt-60, iridium-192 and thulium-170. (The numbers are the atomic mass numbers of the radio-active nuclei.)

There is a continuous reduction in the intensity of radiation emitted from a γ-source as more and more unstable nuclei decay. The rate of decay decreases exponentially with time according to:

$$I_t = I_0 e^{-kt}$$

where I_0 is the initial intensity, I_t is the intensity of radiation at time t, and k is a constant for a particular disintegrating atomic species. An important characteristic of each particular radio-isotope is its *half-life* period. This is the time taken for the intensity of the emitted radiation to fall to one-half of its original value. After two half-life time intervals, the intensity will fall to one-quarter of the original value, after three half-life time intervals the intensity will fall to one-eighth of the original value, and so on.

If the half-life period is T, and I_T is the intensity at time T, then:

$$I_T = \frac{I_0}{2} = I_0 e^{-kT}$$

therefore

$$e^{-kT} = \frac{1}{2}$$

or

$$kT = \ln 2$$

therefore

$$k = \frac{\ln 2}{T}$$

Another characteristic of a γ-ray source is the source strength. The source strength is the number of atomic disintegrations per second and is measured in curies (one curie is 3.7×10^{10} disintegrations per second). The source strength decreases exponentially with time, and the source strength at any given time can be determined using the expression

$$S_t = S_0 e^{-kt}$$

The radiation intensity, usually measured in roentgens per hour at one metre (rhm), is given by source strength (in curies) × radiation output (rhm per curie). The value of radiation output is a constant for any particular isotope.

Another term used in connection with γ-ray sources is *specific activity*. This characteristic, expressed in curies per gramme, is a measure of the degree of concentration of the source.

The characteristics of the commonly used γ-sources for radiography are given in table 6.2.

Table 6.2 Characteristics of γ-ray sources

Source isotope	Half-life period	Photon energy (MeV)	Radiation output (rhm/curie)	Effective penetrating power (mm of steel)
Caesium-137	33 years	0.66	0.39	75
Cobalt-60	5.3 years	1.17, 1.33	1.35	225
Iridium-192	74 days	12 rays from 0.13 to 0.61	0.55	75
Thulium-170	128 days	0.084	0.0045	12 mm aluminium

Commercial radio-active sources are usually metallic in nature, but may be chemical salts or gases adsorbed on carbon. The source is encapsulated in a thin protective covering. This may be a thin sheath of stainless steel or aluminium. By containing the radio-active material in a capsule of this type, it prevents spillage or leakage of the material and reduces the possibility of accidental mis-handling. The encapsulated source is housed in a lead-lined steel container. Two types of container are in general use. In one type the source remains in a fixed position in the centre of the container and the container has a conical plug which may be swung away to allow radiation to emerge. This type of container is sometimes referred to as a 'radioisotope camera'. The other type of container possesses remote-controlled mechanical or pneumatic devices that will open the container and move the source out on a telescopic stalk. When the exposure is completed, the source can be returned to the container and the lid secured by remote control. This second type of container is the more widely used type. The remote-control facility allows the operator to remain at a safe distance from the source. Another advantage is that radiation is emerging from the source at all angles. The source can be located in the centre of a shielded radiation laboratory and a large number of items for inspection, say a production batch of castings, be positioned around the source and all radiographed simultaneously.

6.9 Attenuation of radiation

X-rays and γ-rays interact with the atoms of any medium, including air, which they pass through, and will be attenuated in some way. It is the differential attenuation of radiation by different media which allows radiography to be a useful inspection method. The degree of attenuation is affected by many factors, including the density and structure of the medium and also the type, intensity and photon energy of the radiation.

The intensity of the radiation which emerges from a homogeneous medium decreases exponentially with the thickness of the medium through which it has passed. This can be written as $I = I_0 e^{-\mu t}$ where I is the intensity of the emerging radiation, I_0 is the intensity at the entry face, t is the thickness of the medium and μ is a characteristic of the medium termed the *linear absorption coefficient*. This is not constant for all conditions but varies with the photon energy of the radiation. The absorption coefficient of a material is sometimes expressed as a mass-absorption coefficient, μ/ρ, where ρ is the density of the medium. The absorption coefficient can also be expressed as the effective absorbing area of a single atom. This value, known as the *atomic-absorption coefficient* or *atomic-absorption cross-section*, μ_a, is the linear absorption coefficient divided by the number of atoms per unit volume. It is usually expressed in barns (1 barn = 10^{-22} mm^2).

There are several ways in which a photon of X-radiation or γ-radiation can interact with the atoms of a medium, and the most important of these are the photo-electric effect, Rayleigh scattering, Compton scattering and pair production.

The photo-electric effect is an interaction in which a photon is consumed in breaking the bond between an orbital electron and its atom. Any excess photon energy beyond the bond strength is absorbed as electron kinetic energy. The photo-electric effect is negligible with low atomic number elements and for photon energies exceeding about 100 keV, but for high atomic number elements and for photon energies of up to 2 MeV it accounts for a major portion of total absorption.

Rayleigh scattering is an interaction in which a photon is deflected without any loss of photon energy or release of electrons. The angle of deflection is high for low-energy photons and low for high-energy photons.

Compton scattering is an interaction in which an incident photon causes an orbital electron to be ejected from an atom. Only a portion of the photon energy is used for this purpose and the photon emerges at some scatter angle with a lower energy, that is, of a longer wavelength. This longer wavelength may be in the visible sector of the electro-magnetic spectrum.

Pair production can occur if the incident photon energies exceed 1 MeV. Two lower-energy photons of scattered radiation are emitted for each high-energy photon absorbed.

The total absorption is the sum of the absorption or scattering effects of the four processes listed above. This is termed *narrow beam absorption* and is based on the premise that any photon scattered, even if the scatter angle is small, is considered as a photon absorbed. In practice, with a broad beam, photons which are scattered forwards at small angles are not lost but tend to increase the intensity of the attenuated beam. In other words, the broad beam absorption coefficient of a material is less than the narrow beam absorption coefficient.

The theoretical derivation of absorption coefficients is based on the assumption that the radiation is monochromatic, that is, all photons have the same energies and, hence, wavelength. In practice, an X-ray tube produces 'white' radiation and the effective absorption coefficients are modified by the range of photon energies present in the incident beam.

Empirically derived absorption coefficients are the values used most in exposure time calculations.

The radiation emerging from the material being inspected is composed, then, of direct, but attenuated, radiation and scattered radiation. The ratio of the intensity of the scattered radiation to the intensity of the direct radiation is known as the *scattering factor*. The scattered radiation does not help to reveal characteristics of the testpiece but, rather, detracts from the quality of image possible in that it tends to reduce contrast on the film, obscure detail and reduce overall sensitivity. The scattering factor varies with testpiece thickness, increasing as the thickness increases, and also varies with tube voltage, decreasing as the tube voltage increases (namely, decreasing as photon energies increase) because low photon energy radiation is scattered more than high photon energy radiation (see figure 6.10).

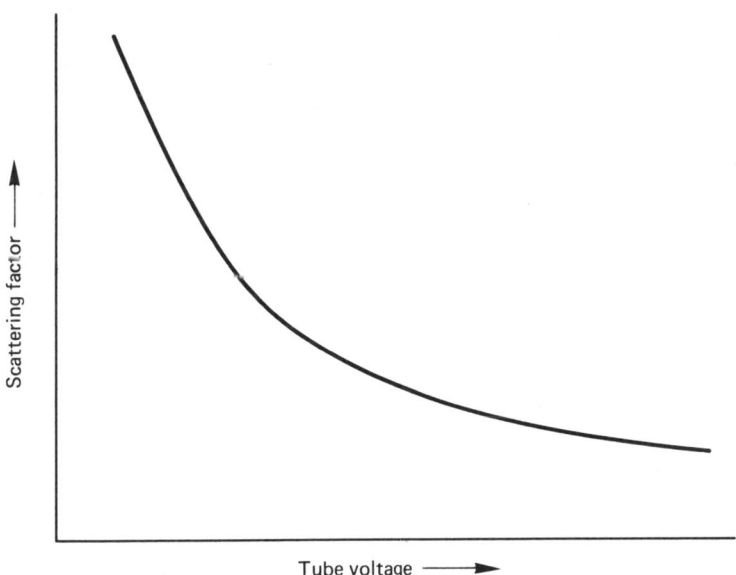

FIGURE 6.10   Variation of scattering factor with tube voltage.

It is desirable then, to use high-energy or short-wavelength radiation (hard X-rays) for good contrast and definition. This can be achieved by either increasing the voltage of the X-ray tube or by using a selective filter to absorb the longer-wavelength portion of the incident radiation. The optimum choice of filter will require consideration of testpiece material and thickness, film type, tube voltage and tube current. A high atomic number element used as a filter will tend to produce a clean image but this would require lengthy exposure times, with consequent cost increase, because it would also cause a major reduction in the intensity of the radiation reaching the testpiece.

Raising the tube voltage can frequently produce high-quality radiographs at a reduced cost. The improved image quality is largely achieved by reducing the proportion of image density due to scattered radiation, which may be of the order of 80 per cent when low voltage, or soft, X-rays are used.

6.10 Radiographic equivalence

The absorption of X-rays and γ-rays by various materials becomes less dependent on composition as the energy of the radiation increases. For example, at 150 kV, 1 mm thickness of lead is equivalent to 14 mm of steel but for photon energies of

1 MeV, then a 1 mm thickness of lead is only equivalent to a 5 mm steel thickness. The absorption equivalence values for some materials are given in table 6.3. Equivalence values are useful when it comes to establishing exposure times for various materials and components.

Table 6.3 Radiographic absorption equivalence for some metals

Material	X-rays (kV)			X-rays (MeV)			γ-rays		
	150	220	400	1	2	4	Cs-137	Co-60	Ir-192
Steel	1.0	1.0	1.0	1.0	1.0	1.0	1.0	1.0	1.0
Magnesium	0.05	0.08	–	–	–	–	–	–	–
Aluminium	0.12	0.18	–	–	–	–	0.35	0.35	0.35
Al–4% Cu alloy	0.16	0.22	–	–	–	–	0.35	0.35	0.35
Titanium	0.45	0.35	–	–	–	–	–	–	–
Copper	1.6	1.4	1.4	–	–	1.3	1.1	1.1	1.1
Zinc	1.4	1.3	1.3	–	–	1.2	1.0	1.0	1.1
Lead	14.0	12.0	–	5.0	2.5	3.0	3.2	2.3	4.0

6.11 Shadow formation, enlargement and distortion

X-rays and γ-rays, like light, propagate in straight lines and the geometric relationships between source, object and film, or screen, determine the main features of the radiographic image which is formed. The image formed on a radiograph is similar to the shadow cast on a screen by an opaque object placed in the light path.

The dimensions of the shadow of an object illuminated from a point source will be larger than the object (see figure 6.11a). This enlargement effect is not of major consequence in radiography because the film or other recording medium is generally placed close behind the testpiece, although with a testpiece of complex shape, some portions of the testpiece may be a relatively long way from the plane of the film and, hence, enlarged. There are some instances in radiography in which enlargement can be used to advantage, and detail that might otherwise be invisible may become visible by means of enlargement.

If a plane of the object being inspected is not parallel to the film or other recording medium, the shape of the image will be distorted (see figure 6.11b). Radiographs of many test objects will show some distortion because many testpieces will be of such shape that they will possess some features that will not be parallel to the plane of the film. In a number of cases the object may contain features that will produce images which overlap one another. This type of behaviour is shown in figure 6.11c.

Most radiation sources are too large to be considered as a point source. When a source of finite size produces a shadow, there will be a certain geometric unsharpness to the shadow which is formed. There will be a portion of the image that will be in shadow for radiation coming from all points on the source. This region of complete shadow is termed the *umbra*. There will also be portions of the image

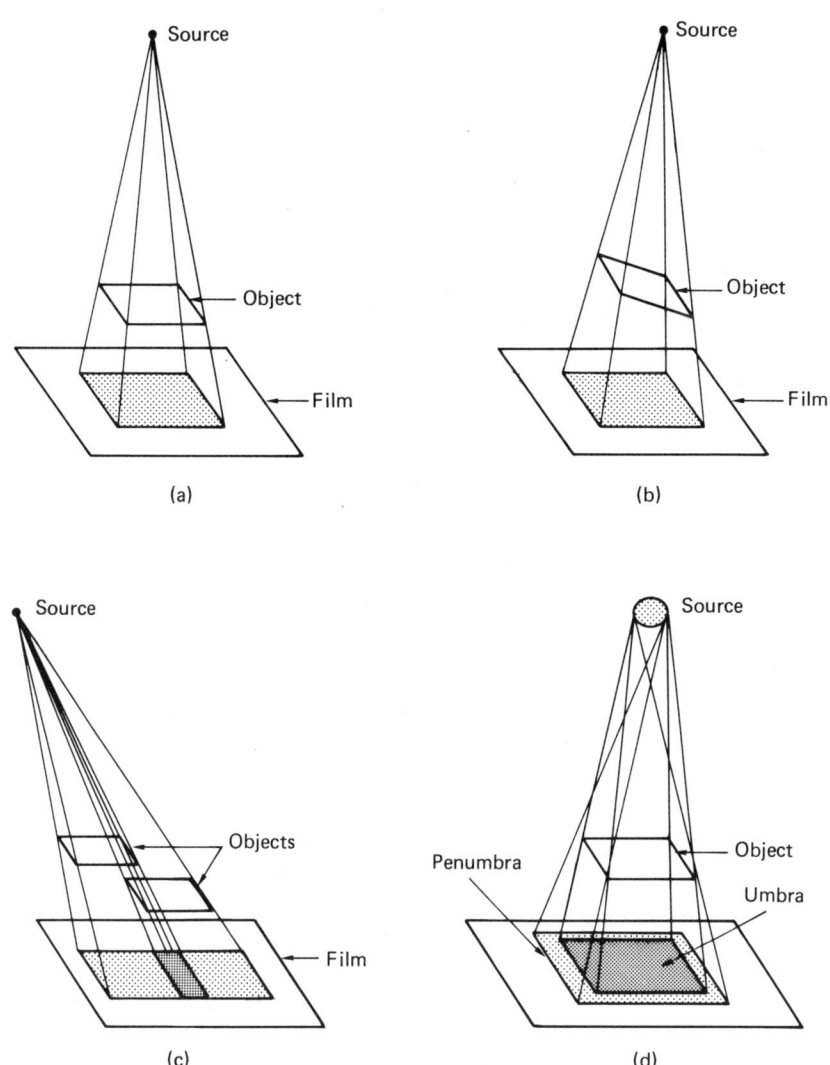

FIGURE 6.11   Geometric features: (a) image enlargement; (b) image distortion; (c) formation of overlapping images; (d) unsharpness of image produced by source of finite size.

which are in shadow for radiation from some portions of the source but not in shadow for all the radiation. This zone of partial shadow is termed the *penumbra* (see figure 6.11d). The size of the penumbra, or the degree of geometric unsharpness, can be reduced by reducing the size of the source of radiation, by reducing the object to film distance, or by increasing the source to object distance.

There is a special technique of high-definition radiography possible utilising an X-ray source with a very small focal spot. The size of the focal spot may be as small as 12 μm in diameter. This gives, in effect, a point source of radiation, and penumbra effects are almost entirely eliminated. This makes it possible to increase significantly the object-to-film distance to give magnification.

In practice, the dimensions of the source are fixed by either the pellet size of an isotope source or the characteristics of the X-ray tube being used, and the object-to-film distance is a minimum with, in many cases, the film being placed in contact with the surface of the object. Thus, the only parameter which can be varied is the source-to-object distance. However, any increase in the source-to-object distance will result in an increase in the exposure time necessary and the radiographer has to compromise between high-definition images with the minimum degree of unsharpness, and costs which are closely linked to the length of exposure time.

6.12 Radiographic film and paper

The film which is used for industrial radiography differs from normal photographic film in a number of respects. The film base is considerably thicker than photographic film, being about 0.018 mm in thickness and it possesses a thin film of emulsion on both sides. Some radiographic film is produced with only one emulsion layer but this type is not in general use. The silver salts in the emulsion are very sensitive to electro-magnetic radiation of the X-ray or γ-ray type but are also sensitive to visible light. A thin coating of emulsion on both sides of the film base effectively increases the speed of the film because the radiation affects both layers almost equally. The film development process is also more effective than would be the case for one thick emulsion layer rather than two thin layers.

X-ray film is of two general types: direct exposure no-screen-type, and screen-type film. Most industrial X-ray film is of the direct exposure type and is available in several grades, each grade with a different combination of film speed and grain size. The film produced for medical radiography is of the screen-type but this grade is used for some industrial applications. The type of application for which medical radiographic film would be used is where a low-power radiation source is involved and where very long exposure times would be needed were direct exposure film to be used. The emulsion on medical screen-type film is more sensitive to visible light than to X-radiation, and is particularly sensitive to the wavelengths emitted by the fluorescent screens with which they are used.

X-ray images can also be recorded on radiographic paper. The paper only has an emulsion layer on one side. The paper may be used in the same manner as direct exposure film, but it is generally used in connection with fluorescent screens because it allows for shorter exposure times and gives a good contrast image. The emulsion on a radiographic paper usually contains developer chemicals which are activated by dipping the exposed paper in an alkaline solution. Processing is very rapid but the image will show some deterioration after a period of 8 to 10 weeks. If it is required to store a paper radiograph for a longer period than that it should be fixed, washed and dried in the conventional manner. Radiographic paper is frequently used instead of film in some process-control inspection, by virtue of its speed, convenience and lower cost.

6.13 Xeroradiography

In this system the latent radiographic image is produced on a metal plate which has a thin coating of selenium. The plate is given an overall but even charge of static electricity. A charged plate of this type is highly sensitive to light and also X-radiation and must be kept in a light-tight holder similar to a film cassette. When X-radiation impinges on the plate it will differentially discharge the surface of the plate in proportion to the amount of radiation received at different portions of the plate, thus producing an electrostatic latent image.

The exposed plate is developed by subjecting it, in the absence of light, to a cloud of extremely fine plastic powder, or toner. The particles of the toner powder are charged electrostatically, but with a charge of opposite sign to the charge on the plate. The toner powder particles are attracted to the charged portions of the plate, producing a radiographic image. This image is not permanent as the powder particles are merely held in place by the electrostatic charge. A permanent image can be made by placing a piece of treated paper on the plate. The powder is transferred to the paper where it can be fixed in position using heat. Xeroradiographic images can normally show excellent detail.

6.14 Fluoroscopy

In fluoroscopy the X-ray image is not recorded on film but the radiation, after passing through the object under inspection, impinges on a screen made up of crystals of a fluorescent compound bonded on a thin base. The crystals fluoresce when X-radiation strikes them and the intensity of the fluorescence produced is greatest in those parts receiving the most radiation. A sketch of the principle of fluoroscopy is shown in figure 6.12. The appearance of the fluoroscope image is the reverse of that seen on a developed radiograph and a defect such as porosity which would show as dark spots on a radiograph would appear as brighter spots on a fluorescent screen.

The major advantage of fluoroscopy is that an image can be seen in 'real-time', and it is a useful inspection technique for the inspection of large-scale production items.

A fluoroscopic image tends to be fairly dim and should be viewed in a room with a low level of background illumination. Image amplification can be used to give a brighter image and, hence, improve the ease with which fine detail can be

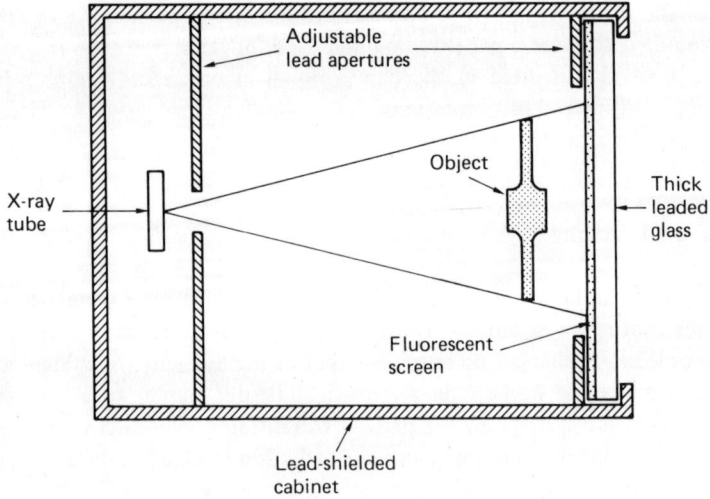

FIGURE 6.12   Principle of operation of a fluoroscope.

seen. The brighter image formed after amplification is smaller than the original screen image but is bright enough to be enlarged by an optical lens system. Alternatively, the bright image could be viewed remote from the X-ray cabinet using a closed circuit TV system.

A development, but one which still comes under the general heading of fluoroscopy, is to use a solid-state imaging panel in place of a fluorescent screen. By this means, images of high quality can be displayed on a TV monitor screen.

An inspection system for the examination of die cast aluminium car wheels is shown in figure 6.13. Figure 6.13a shows wheels on a conveyor entering a shielded X-ray cabinet. The system incorporates a 160 kV constant potential X-ray unit and an electronic image amplification unit. The image can be viewed on a TV monitor which is remote from the radiation area (figure 6.13b). The operator in the viewing area has full control of the X-ray unit and wheel movement.

6.15 Exposure factors

There are very many factors which govern the formation of a satisfactory image on radiographic film. When it is desired to radiograph some object or component, the composition, density and dimensions of the object will largely determine the quality of the radiation which will be used. The source must produce radiation which will be sufficiently penetrating for the type and thickness of material to be inspected, and so X-ray tube voltage or type of γ-radiation source will be selected accordingly. The selection of a particular grade of radiographic film will be made

(a)

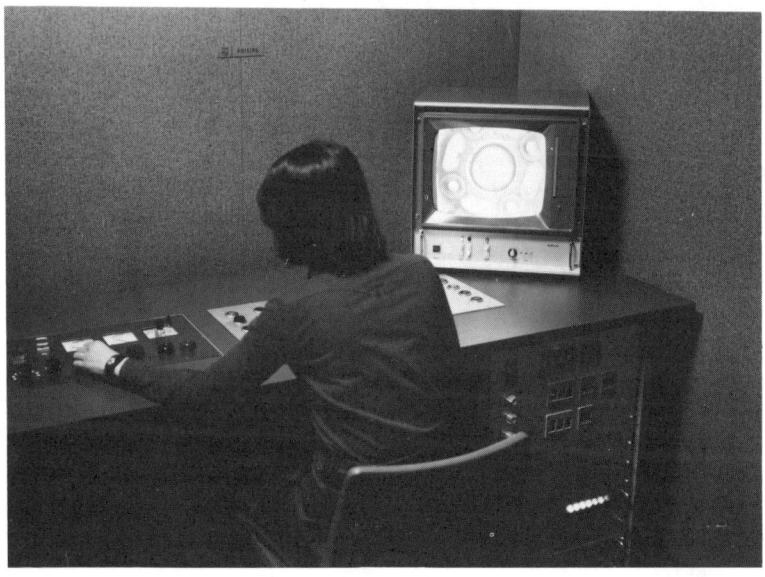

(b)

FIGURE 6.13   Fluoroscopy inspection system: (a) wheels entering shielded X-ray
              cabinet: (b) remote viewing room, showing image of wheel on the
              TV monitor and the control panel (courtesy of Wells–Krautkramer
              Ltd).

on the basis of its sensitivity to the variations of radiation intensity which are expected after transmission through the object. However, the amount of radiation expected to reach the film will be affected by several other factors, including the intensity of the incident radiation (governed by the X-ray tube current or the strength of a γ-source in curies), the source-to-film distance and the exposure time. The correct exposure for a particular application may be determined by a process of trial and error or by using an aid such as an exposure chart which relates to a specific grade of film. A typical exposure chart is shown in figure 6.14. It will be noticed that one axis of the chart is marked in milliamp-seconds. The intensity of radiation emitted from an X-ray tube at any particular tube voltage is proportional to the tube current and so, if a radiograph were made with an exposure time of 8 seconds at a tube current of 20 mA, an equivalent radiograph could be obtained, assuming tube voltage to be constant, with a 16 second exposure at 10 mA or a 10 second exposure at 16 mA.

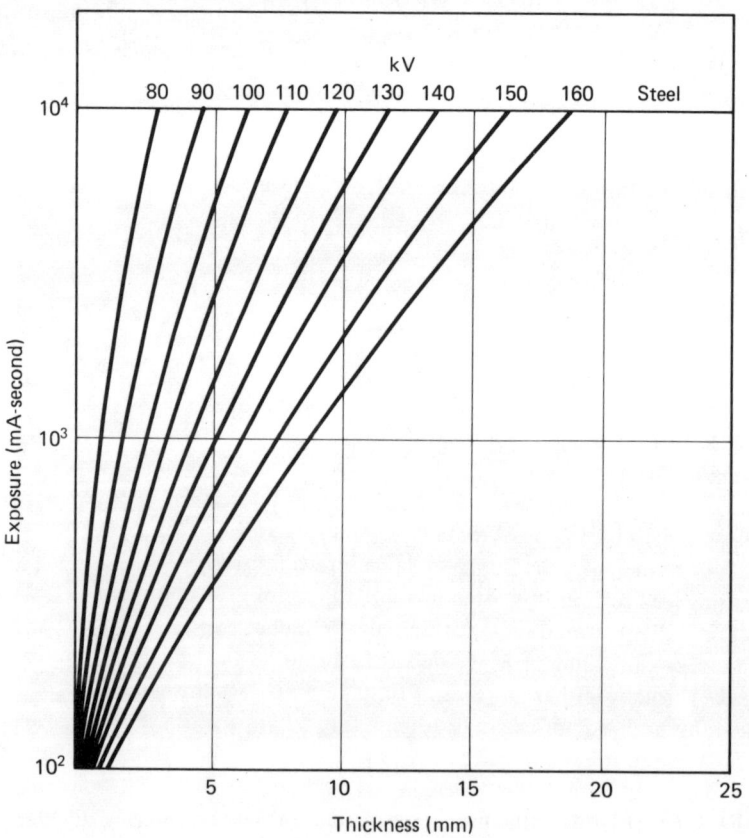

FIGURE 6.14   Typical X-ray exposure chart for steel.

The intensity of radiation at any point diminishes as the distance from the source increases, according to the inverse square law. This can be written as $IL^2 =$ constant, where I is the intensity and L is the distance from the source. An exposure chart, as shown in figure 6.14, gives data for one particular source-to-film distance, usually one metre. If the source-to-film distance differs from this, then the exposure time has to be adjusted accordingly. For example, if the exposure chart recommends an exposure of, say, 200 mA-second, at a source-to-film distance of 1 metre, then for an equivalent radiograph the exposure would need to be 200×4 mA-second at 2 metres, 200×2.25 mA-second at 1.5 metres, or 200×0.64 mA-second at 0.8 metres. Further modifications to the figures given by an exposure chart will need to be made if a filter is used or if screens are used.

It should be borne in mind, though, that these exposure charts are, at best, only a guide. They are only strictly accurate for testpieces of uniform thickness and for one specific X-ray installation because each type of X-ray unit differs from one another and indeed each X-ray installation is unique.

Exposure charts for γ-ray sources are constructed in a similar manner to those for X-rays but instead of the exposures being quoted in millamp-seconds they are quoted in curie-minutes or curie-hours. The source strength must also be known in order to use a γ-ray exposure chart. As stated earlier, the strength of an isotope source decreases exponentially with time. Source manufacturers state the source strength on a given date, and from this the strength at any other time can be calculated.

6.16 Radiographic screens

Screens are often used to improve contrast and to increase the density of the radiographic image on film. These screens, which are placed in close contact with the film during exposure, may be either metallic or fluorescent intensifying screens. The most widely used type of metallic screen is the lead screen. Sometimes a combination of both types, the fluoro-metallic screen is used.

By using lead screens it is often possible to improve the contrast in a radiograph by filtering out scattered radiation and also to reduce exposure times. The screens used are very thin films of lead bonded to a thin card or plastic. The thickness of the lead is usually either 0.125 mm or 0.25 mm, and the screens are placed in contact with each side of the radiographic film within the film cassette.

As mentioned in section 6.9 a heavy metal, such as lead, will absorb low-energy (longer-wavelength) radiation much more readily than it will absorb high-energy radiation. The scattered radiation from a testpiece is always of a lower energy than the incident radiation and will be almost entirely absorbed by the screen, while much of the high-energy incident radiation will pass through the lead screen filter. Scattered radiation, 'back scatter', will also be formed from the table or floor on which the film holder is placed. The screen placed behind the film will absorb this back scatter.

One of the effects when high-energy radiation strikes a metal such as lead is an interaction resulting in the emission of electrons (see section 6.9). The electrons emitted from the lead screen material will affect the radiographic film emulsion and give greater developed film densities than would be achieved without screens. This image intensifying effect also gives enhanced contrast which improves the ability to observe small flaws. Plate 1 shows portions of two radiographs taken of the same brass casting. Both radiographs were taken with Kodak Industrex CX film at 180 kV. In each case the source-to-film distance was 700 mm, the tube current was 8 mA and the exposure time was 2 minutes. No screens were used for exposure (a) but a pair of 0.125 mm lead screens was used for exposure (b).

Because lead screens give both a filtration effect and an intensifying effect, there will be some combination of incident wave energy and material thickness at which these effects just balance one another and there is no advantage in using a screen. With steel this point comes at a metal thickness of about 6 mm and a tube voltage of about 130 kV. Below these values the filtration effect is the dominant one, resulting in longer exposure times. For light metals such as aluminium, the minimum thickness at which lead screens are advantageous will be greater than that for steel.

Fluorescent intensifying screens can also be used to improve the efficiency of radiography. The most widely used screen material is calcium tungstate crystals on a thin card base. These screens are very sensitive to X-rays and will fluoresce, giving visible light in the blue sector of the spectrum, and can intensify the film image by a factor of up to 100. However, these screens do not reduce scatter and, in consequence, the image quality is not as good as that obtainable using lead screens. Fluorescent screens are much less sensitive to γ-radiation than to X-rays and will give an intensification of some 20 to 40 times, but the unsharpness of fluorescent screen images, coupled with the generally low contrast of γ-radiographs, means that this type of screen is rarely used in connection with γ-radiography.

Fluoro-metallic screens combine the advantages of both the lead and fluorescent screens. They comprise a combination of a fluorescent layer and a thin sheet of lead and they are mounted with the fluorescent side in contact with the film. The lead filters out scattered radiation and the fluorescent portion intensifies the film image. A good image quality is obtainable and exposure times can be reduced considerably compared with those which would be necessary for film with metallic lead screens only.

6.17 Identification markers and Image Quality Indicators

It is necessary to identify radiographs so that a particular film can always be related to a particular testpiece, or portion of a testpiece, and identification markers are used for this purpose. In addition to identification, each radiograph should contain some means of assessing the quality or sensitivity of the image. This is achieved using devices termed *Image Quality Indicators* (IQIs) or *penetrameters*.

Identification markers are made of lead or lead alloy, usually in the form of letters and numbers. These are then attached to either the testpiece or the film cassette by means of adhesive tape during the setting-up process. The markers should be placed in such a way that they will not obscure any portion of the testpiece because the shadows of the dense metal characters would mask coincident defects.

Several Image Quality Indicators or penetrameters of different designs have been devised by the various standards organisations and they are generally made in the form of steps or wires of varying thickness and made in the same or a similar material to that being inspected.

The relevant British Standard (BS 3971) covers both step-hole and wire IQI types, and these are shown in figure 6.15. Step-hole IQIs to BS 3971 may be machined as a series of steps from a single plate or may be composed of a series of separate plaques mounted on plastic or rubber. Table 6.4 gives the step and hole dimensions.

Table 6.4 Step and hole dimensions for IQIs to BS 3971

Step no.	Step thickness and hole dia. (mm)	Step no.	Step thickness and hole dia. (mm)
1	0.125	10	1.00
2	0.160	11	1.25
3	0.200	12	1.60
4	0.250	13	2.00
5	0.320	14	2.50
6	0.400	15	3.20
7	0.500	16	4.00
8	0.630	17	5.00
9	0.800	18	6.30

Each step contains one or two holes of diameter equal to the step thickness with steps numbered 1 to 8 having two holes and steps numbered 9 to 18 having one hole. A straight six-stage step-hole IQI is shown in figure 6.15a while figure 6.15b shows the model B hexagonal IQI. Identification markers must be put on to indicate the type of material and the thickness range of the IQI. In figure 6.15a the coding 8 AL 13 indicates that the IQI is an aluminium with the thinnest step being No. 8 (0.630 mm) and the thickest step being No. 13 (2.00 mm), while in figure 6.15b the coding 7 FE 12 means that the IQI is in steel with step thicknesses ranging from No. 7 to No. 12. Figure 6.15c shows an example of the wire type IQI conforming to BS 3971. A series of wires, each of 30 mm length, are laid parallel and set 5 mm apart within an optically transparent material such as polythene. The lead identification markers are also encased within the polythene. In figure 6.15c the coding 9 CU 15 means that there are seven copper wires ranging in diameter from wire No. 9 (0.200 mm dia.) to wire No. 15 (0.80 mm dia.). The various wire diameters are given in table 6.5.

Table 6.5   IQI wire diameters (BS 3971)

Wire No.	Diameter (mm)	Wire No.	Diameter (mm)	Wire No.	Diameter (mm)
1	0.032	8	0.160	15	0.80
2	0.040	9	0.200	16	1.00
3	0.050	10	0.250	17	1.25
4	0.063	11	0.320	18	1.60
5	0.080	12	0.400	19	2.00
6	0.100	13	0.500	20	2.50
7	0.125	14	0.630	21	3.20

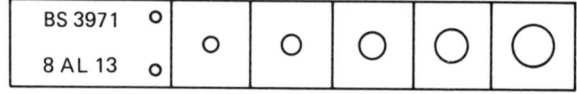

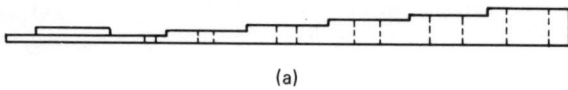

(a)

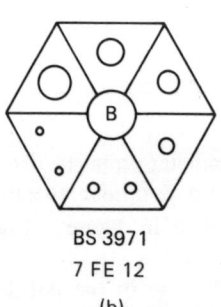

(b)   (c)

FIGURE 6.15   Penetrameters to BS 3971: (a) step-hole type; (b) hexagon model B; (c) wire type.

Another type of wire penetrameter which is very widely used is the one to the German standard 54109. This is the DIN Type 62 (Deutsche Industrie Norm) which is illustrated in figure 6.16a. The dimensions of the wires in the DIN 62 penetrameter are given in table 6.6. The coding, in lead characters, below the wires identifies the wire sizes. In figure 6.16a there are seven wires ranging in size from No. 10 (0.4 mm) to No. 16 (0.10 mm).

Table 6.6 Wire sizes in DIN 62 penetrameter

Wire No.	Diameter (mm)	Wire No.	Diameter (mm)	Wire No.	Diameter (mm)
1	3.2	7	0.8	12	0.25
2	2.5	8	0.63	13	0.2
3	2.0	9	0.5	14	0.16
4	1.6	10	0.4	15	0.125
5	1.25	11	0.32	16	0.1
6	1.0				

The types of penetrameter to meet US specifications differ from those specified in British and European standards. Plaque-type penetrameters are used, as illustrated in figure 6.16b and c.

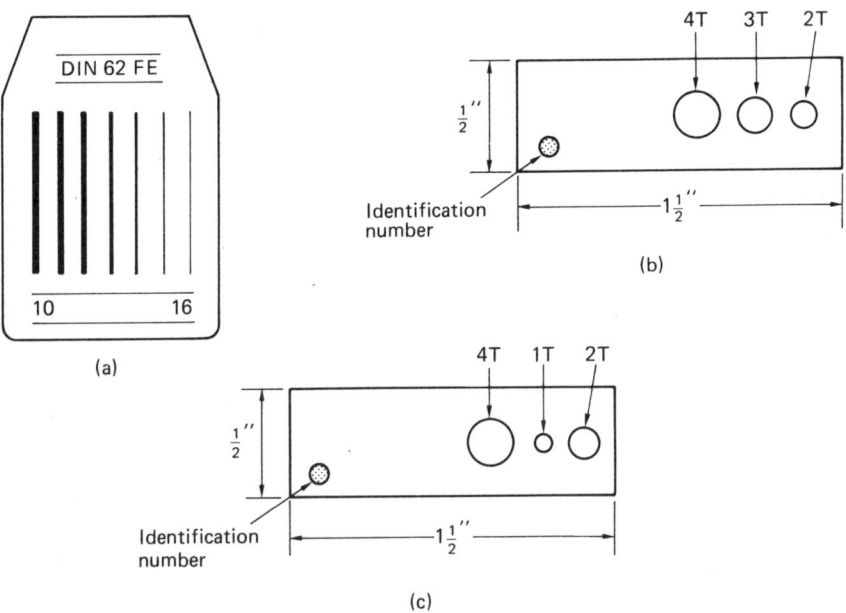

FIGURE 6.16 Types of penetrameter: (a) DIN type 62 wire penetrameter; (b) ASME plaque penetrameter (USA); (c) ASTM (Code E 142-68) plaque penetrameter (USA).

To satisfy the ASME standard, a penetrameter as shown in figure 6.16b is used. This is a thin plaque of a thickness (T) equal to 2 per cent of the testpiece thickness. It possesses three holes, these having diameters equal to twice, three times and four times the penetrameter thickness respectively, with a minimum hole diameter size of 0.068 inch. If the testpiece thickness is greater than 2.5 inches, the size of the penetrameter is increased to 2.5 inches × 1 inch. The ASTM penetrameter (figure 6.16c) is very similar. The identification number of an ASME or ASTM penetrameter is its thickness in thousandths of an inch. The three holes in the ASTM type have diameters equal to once, twice and four times the thickness of the penetrameter subject to minimum diameters of 0.010 inch and 0.020 inch and 0.040 inch respectively.

Using the British or German type penetrameters the image quality, or sensitivity, is expressed as a percentage. The sensitivity is the thickness of the thinnest wire or step or hole visible in the developed radiograph, expressed as a percentage of the thickness of the testpiece. With the ASME standard, where the thickness of the penetrameter is 2 per cent of the testpiece thickness, if all three holes are visible on the developed film the sensitivity is equal to 2 per cent. In the ASTM system, image quality is rated by code symbols, 1–1T, 1–2T, 2–1T, and so on. The first figure in the code is the thickness of the penetrameter as a percentage of the testpiece thickness and the last two characters represent the smallest hole size visible in the developed radiograph. The equivalent percentage sensitivities of these image qualities are given in table 6.7.

Table 6.7 Sensitivities of ASTM image quality levels

Image quality level	Penetrameter thickness (per cent of testpiece thickness)	Smallest visible hole size	Equivalent sensitivity (per cent)
1–1T	1	1T	0.7
1–2T	1	2T	1.0
2–1T	2	1T	1.4
2–2T	2	2T	2.0
2–4T	2	4T	2.8
4–2T	4	2T	4.0

The placement of penetrameters is important. They should be placed on the source side of the testpiece and at the edge of the area, namely in the outer zone of the radiation beam with the thinnest step or wire being outermost.

6.18 Inspection of simple shapes

It is usually the best practice to direct the radiation at right angles to a surface in such a way as to pass through a minimum thickness of material. In this way exposure times will be at a minimum. However, if plane defects such as cracks are suspected within the component the radiation should be directed parallel to the expected crack direction irrespective of testpiece thickness in that direction.

The simplest shapes are flat plates and the radiation should be normal to the surface of the plate. When large areas are to be inspected, this should be done by a series of overlapping exposures. It is often more economical to radiograph in this way. A single large radiograph would require a large source-to-film distance to avoid distortion and this would result in long exposure times, whereas relatively short source-to-film distances can be employed for a series of smaller, but over-lapping, exposures.

Curved plates can be examined in a manner similar to that for flat plates but for the best results the film should conform to the shape of the plate. This is achieved by placing the radiographic film in a flexible light shield which can be secured to the plate surface by means of either magnetic clamps or adhesive tape. Figure 6.17 shows an operator setting up for the examination of the longitudinal weld in a section of a pressure vessel. The X-ray set is a Philips 420 kV unit and the two 210 kV high-voltage generators to power this unit are visible in the background. Again, to avoid image distortion, several overlapping exposures are necessary to examine the whole of this weld.

FIGURE 6.17   Setting-up for radiography of a longitudinal weld in a pressure vessel using a Philips 420 kV X-ray unit (courtesy of Wells–Krautkramer Ltd).

Some X-ray sets, with a rod anode, are capable of emitting a panoramic X-ray beam through a full 360° circumferential tube window. This type of unit is ideally suited for the inspection of circumferential welds in cylindrical pressure vessels or large bore pipes. The X-ray tube is positioned on the centre-line of the vessel or pipe and the film is wrapped around the outside of the structure. This is shown in figure 6.18. A radiograph of a section of a circumferential weld in a large diameter pipe is shown in Plate 2.

FIGURE 6.18   Philips 300 kV radial panoramic X-ray unit used to examine a circumferential weld in a large vessel. The film is wound in a continuous length on the outside of the vessel and held in place by magnetic clamps (courtesy of Wells–Krautkdramer Ltd).

This panoramic technique can also be used for the simultaneous inspection of several small components. The components can be arranged around the X-ray unit and a separate film placed behind each object.

Cylindrical shapes can pose problems for the radiographer. They may either be inspected by a longitudinal view, but this is generally only satisfactory for short but large diameter cylinders, or by a transverse view which is often only satisfactory for relatively small diameter cylinders. Considering a diametral plane of a cylinder, it varies in thickness from zero at the edges to a maximum at the centre. Thus, an exposure time which would be suitable for the central portion of a cylinder would result in over-exposure at the edges. Edge definition is relatively good for light metal cylinders with diameters of less than about 50 mm and for

PLATES

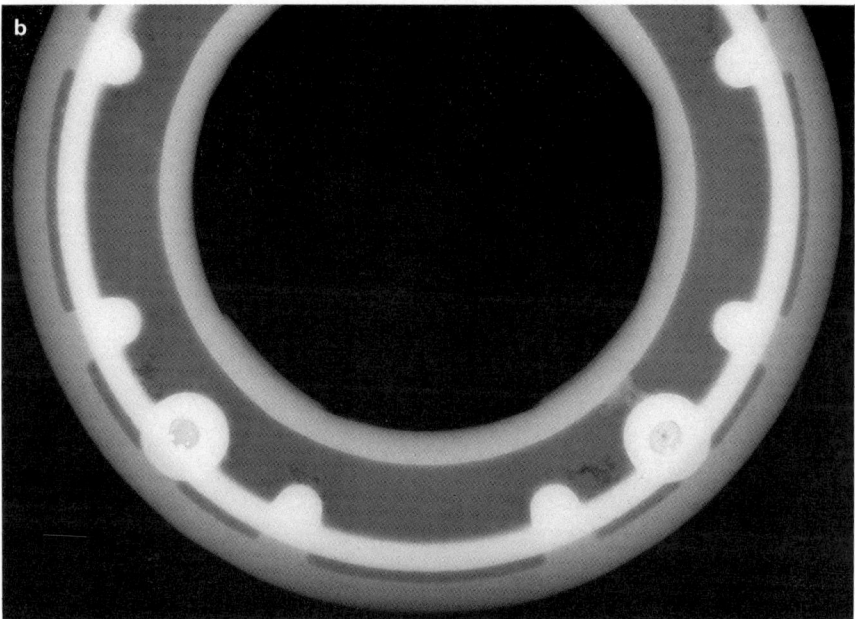

Plate 1 Radiographs of a brass casting with porosity zones and showing the effectiveness of lead screens. (a) Direct exposure. 2 minutes at 8 mA and 180 kV. (b) Exposure with pair of 0.125 mm lead screens for 2 minutes at 8 mA and 180 kV (courtesy of Kodak Ltd)

Plate 2 Radiographs of a circumferential weld in a 36 inch diameter high-pressure gas pipe using the panoramic technique. The number markers indicate distances, in cm, from a datum point. Exposure 1½ minutes at 6 mA and 180 kV with s.f.d. of 450 mm. Industrex AX lead pack film used (courtesy of Kodak Ltd)

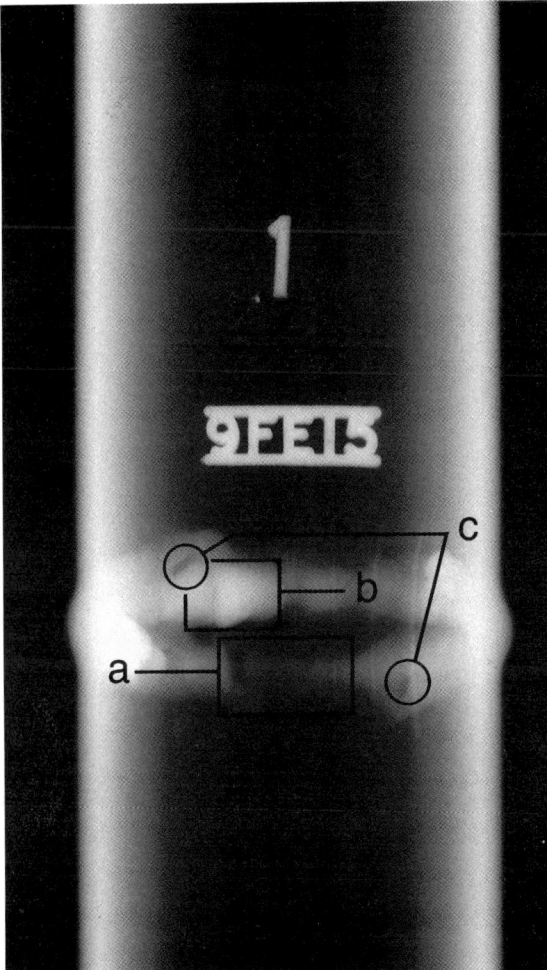

Plate 3 Radiograph of weld in steel pipe of 60 mm o.d. showing (a) lack of penetration, (b) excess penetration and (c) inclusions. Wires 13 to 15 of the penetrameter (to BS 3971) are visible. Exposure 2 minutes at 6 mA and 170 kV. Industrex AX film with lead screens (courtesy of Kodak Ltd)

Plate 4 Radiograph of portion of an aircraft control surface. Exposure 30 seconds at 4 mA and 60 kV. Kodak AX film with no screens (courtesy of Kodak Ltd)

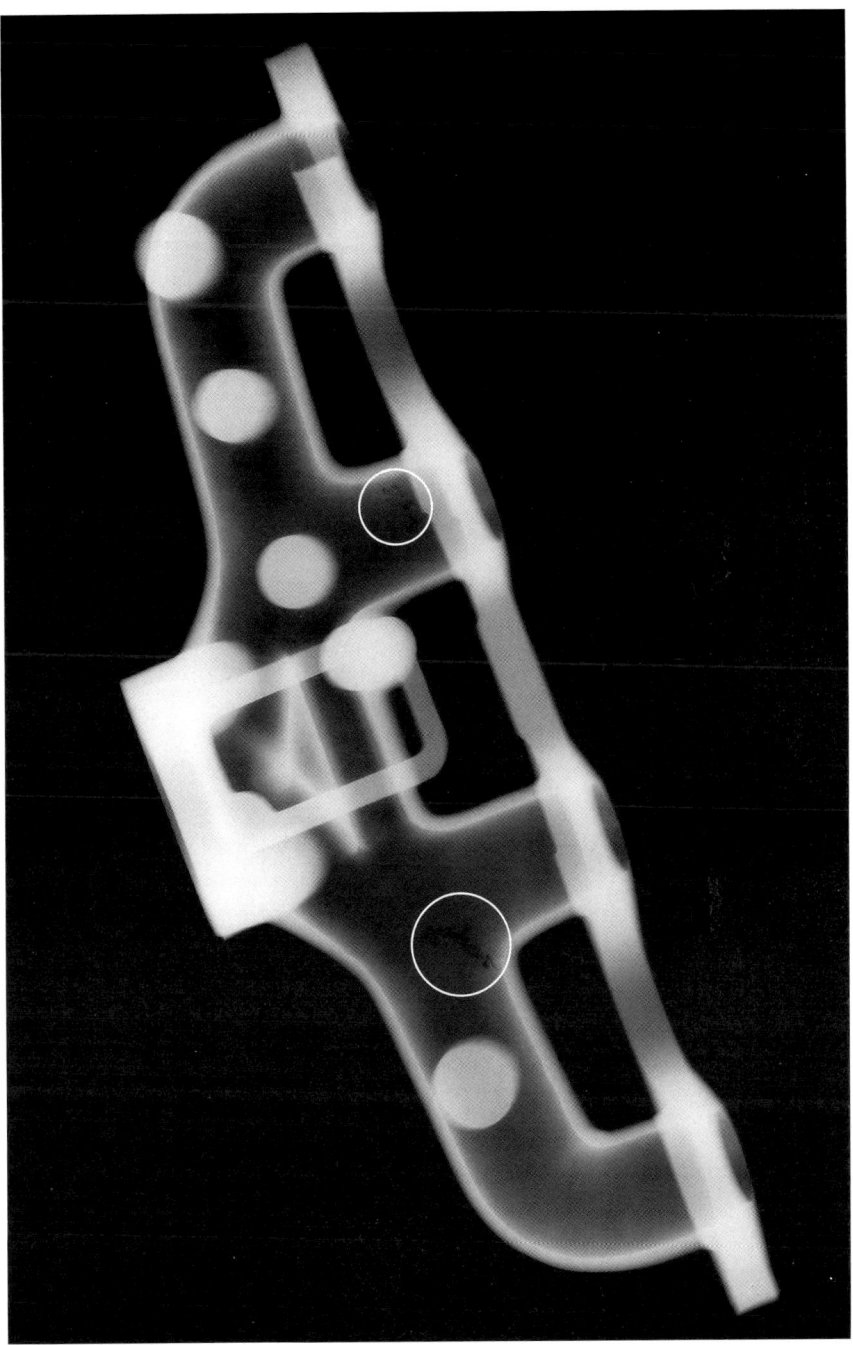

Plate 5 Radiograph of engine manifold. Shrinkage cavities are encircled. Exposure 30 seconds at 10 mA and 80 kV using MX film (courtesy of Kodak Ltd)

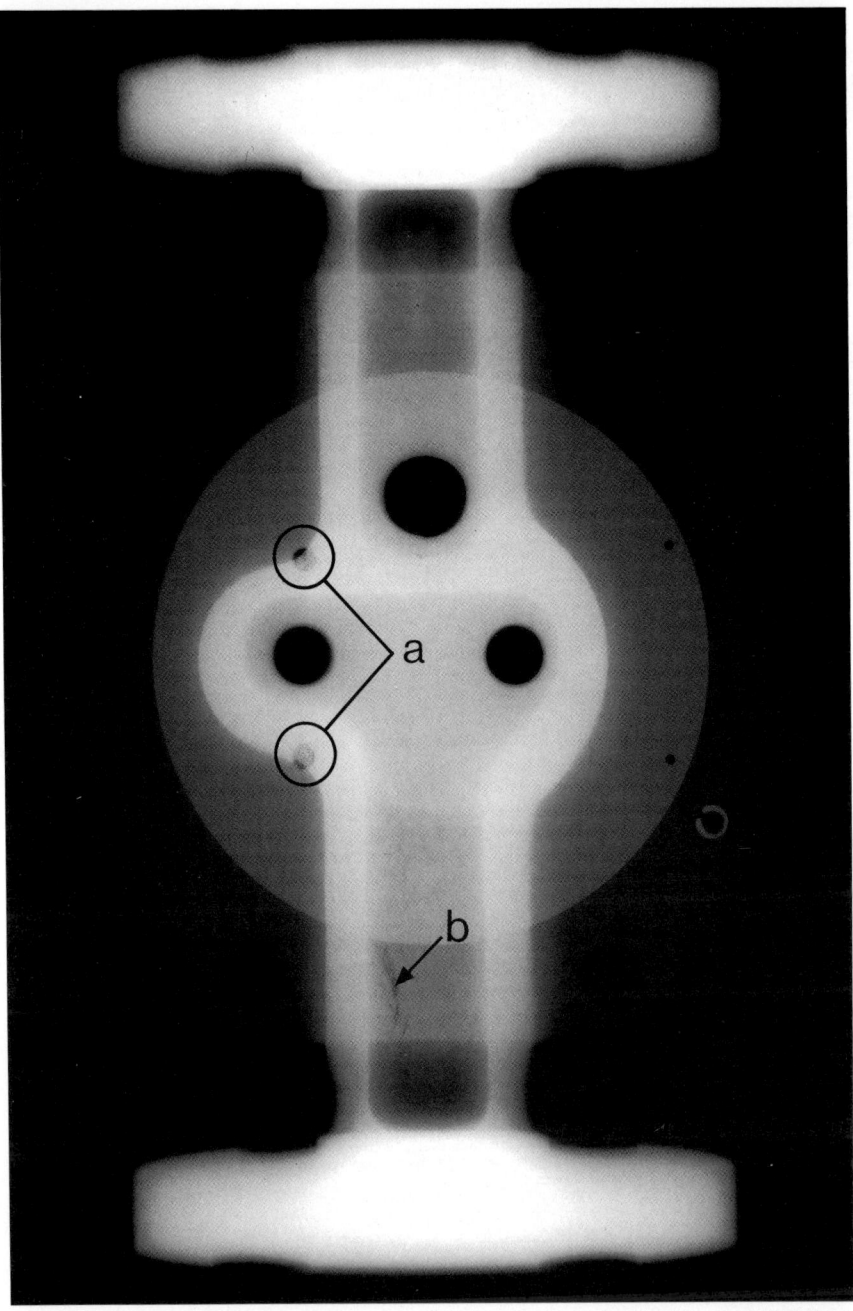

Plate 6 Steel casting showing (a) shrinkage porosity and (b) shrinkage cracks. 250 kV; exposure 45 seconds at 5 mA, and with s.f.d. of 1m. 0.25 mm lead filter used (courtesy of Kodak Ltd)

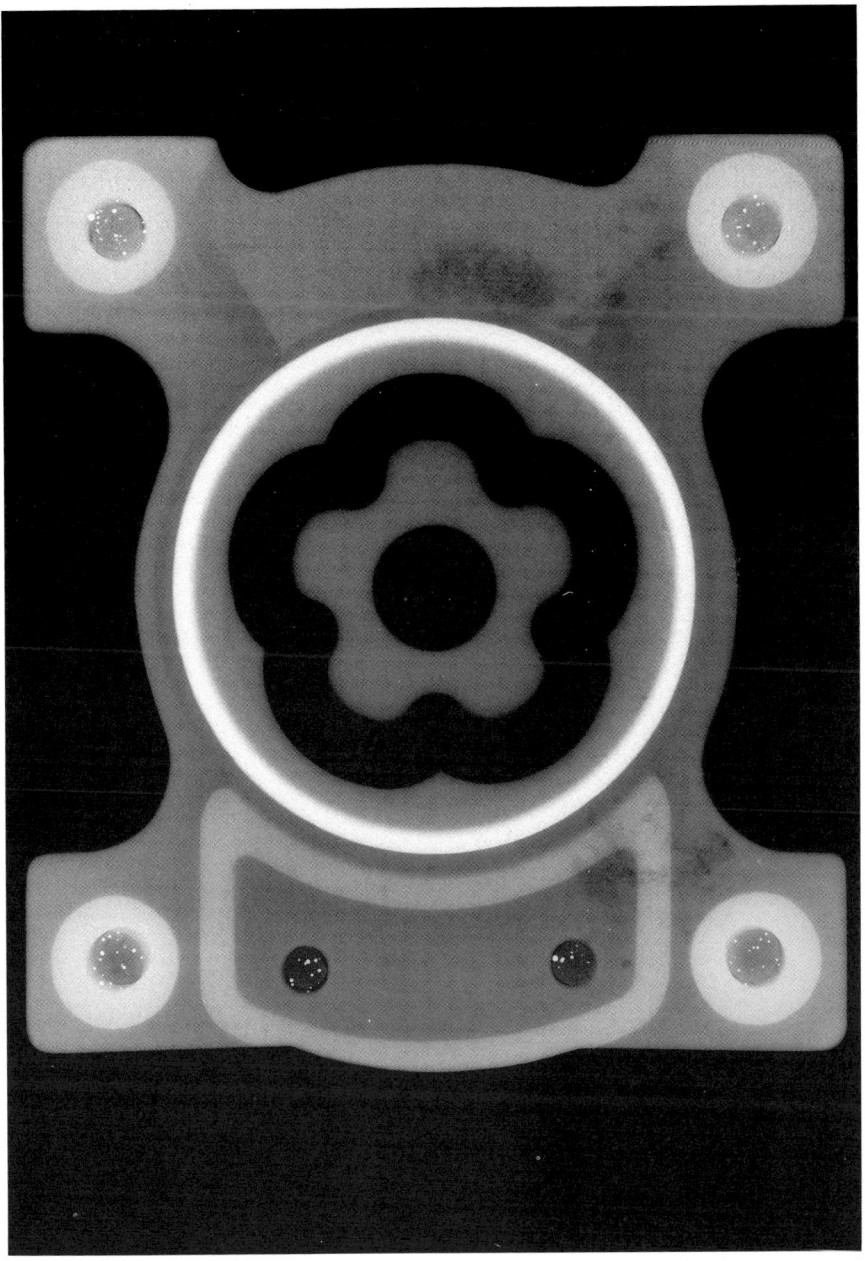

Plate 7 Areas of spongy porosity in a steel casting. Exposure 60 seconds at 5 mA and 200 kV with s.f.d. of 1 m. Industrex AX film with 0.125 mm lead screens used (courtesy of Kodak Ltd)

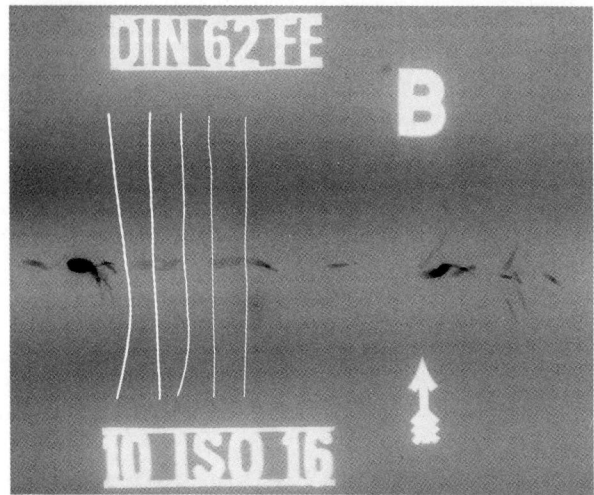

Plate 8 Radiograph of butt weld in 13 mm thick stainless steel showing entrapped argon pockets and DIN type 62 penetrameter. Exposure 60 seconds at 10 mA and 140 kV with 0.125 mm lead screens (courtesy of Kodak Ltd)

Plate 9 Radiograph of release mechanism for a Martin Baker aircraft ejector seat showing correctness of the assembly. Exposure 30 seconds at 5 mA and 250 kV using 0.25 mm lead filter with s.f.d. of 1 m. MX film with 0.125 mm lead screens used (courtesy of Kodak Ltd)

cylinders of steel or other dense metals if the diameter does not exceed a value of about 25 mm. For cylinders with larger diameters than these, there are several techniques which may be used. One method would be to make two exposures using a different tube voltage for each. The exposure at the lower tube voltage would give good edge definition and the exposure at the higher tube voltage would give a good radiograph of the central portion of the cylinder. A second technique would be to make two exposures using the same tube voltage with the cylinder rotated through 90° between the first and second exposure. Alternatively, equalising methods can be used in which the outer edges of the cylinder, where the section is thinnest, are built up to present a greater radiographic density. This can be achieved by mounting the cylinder in a close-fitting cradle or by using a liquid absorbing medium, as shown in figure 6.19. In both cases the absorber is in close contact with the cylinder around one-half of the circumference.

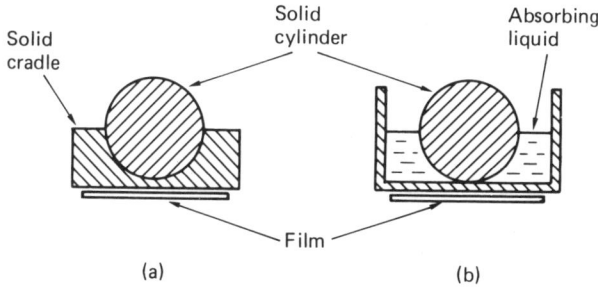

FIGURE 6.19   Equalisation of radiographic density for examination of solid cylinders: (a) use of solid cradle; (b) use of liquid absorber.

There are fewer problems posed by tubular sections than by solid cylinders, because the total variation in thickness across a diametral plane is much less. However, the maximum effective thickness is now at the edges, rather than at the centre. A method which is frequently used for the inspection of circumferential welds in tubes is to direct the X-ray beam at a slight angle from the normal so that the weld area appears on the radiograph as an ellipse. Plate 3 shows a radiograph of a circumferential weld in a pipe of 63 mm outside diameter. The flaws visible include lack of penetration, excess penetration and inclusions.

6.19 Inspection of complex shapes

The inspection of complex shapes often requires multiple exposures from different viewing directions. The selection of each view will depend on the shape of the component and also the likely orientation of anticipated flaws. There are certain guidelines which can be followed in the choice of X-ray view and one is that there should not be great variations of material thickness in the view direction. The view should be such that the radiographic image be of a relatively simple shape to aid interpretation, and the view should also be selected to give minimum unsharpness. The radiograph in Plate 4 is of a complex assembly, namely part of a control surface of an aircraft.

6.20 Viewing and interpretation of radiographs

A radiograph is valueless unless the developed image can be sensibly interpreted, and for correct interpretation needs a person who possesses a considerable amount of knowledge, skill and experience. The interpreter, therefore, needs to have a thorough knowledge of the principles of radiography and to be fully aware of the capabilities and the limitations of the techniques and equipment. In addition, the interpreter should have knowledge of the components to be inspected and the variables in the manufacturing processes which may give rise to defects. For example, in the inspection of castings, it would be beneficial if the interpreter is aware of the way in which defects such as gas porosity, shrinkage and cold shuts can occur and the most likely areas in the particular casting where they may be found.

The radiographic interpreter is looking for changes in image density in the radiograph. Density changes may be caused by one of three factors, namely a change in the thickness of the testpiece, including visible surface indentations or protuberances, internal flaws within the component, and density changes which may be induced by faulty processing, mis-handling or bad film storage conditions, and it is important that the interpreter can assess the nature and cause of each density difference observed.

The conditions in which radiographs are viewed, therefore, are highly important, and the film should be correctly illuminated by means of a purpose-built light source which will give good illumination without glare or dazzle. The radiograph should be viewed in a darkened room so that there will be no light reflections from the surface of the film and the image is seen solely by means of light transmitted through the film. Viewing in poor conditions will cause rapid onset of eye fatigue and so it is also important that the interpreter is in a comfortable position and has no undue distractions. Ultimately, the efficiency of flaw detection is determined by the skill and experience of the interpreter and a highly experienced radiograph interpreter may locate defect indications which could be missed by a less experienced person.

The radiographs shown in Plates 5, 6 and 7 are of various castings showing defects. Plate 5 is a radiograph of a light alloy engine manifold casting, showing a large shrinkage cavity. Plate 6 is a radiograph of a steel casting. Shrinkage has occurred in the thick central section. The casting shown in Plate 7 demonstrates several areas of spongy porosity. This is a sand casting in steel with a maximum section thickness of 15 mm and the radiograph was taken at 200 kV with an exposure of 60 seconds at 5 mA with a source-to-film distance of 1 m. Industrex AX film with 0.125 mm lead screens was used.

A radiograph of a defective weld is shown in Plate 8, which shows an argon arc butt weld between two plates of austenitic stainless steel of 13 mm thickness. The defects are mainly gas pockets of entrapped argon.

An example of the use of radiography to check for the correctness of assembly is shown in Plate 9. This radiograph is of the release mechanism for a Martin Baker aircraft ejector seat incorporating a time delay mechanism and aneroid barometer for height release.

6.21 The radiation hazard

X-ray and γ-radiation can cause damage to body tissue and blood, but any damage caused is not immediately apparent. The effects of any small doses of radiation received over a period of time are cumulative, and so all workers who may be exposed to even small quantities of radiation should have a periodic blood count and medical examination.

Strict regulations cover the use of X-rays and γ-rays, and the quantity of radiation that workers may be exposed to. The unit of quantity of X-rays or γ-radiation is the roentgen (see section 6.6) which is based on the amount of ionisation caused in a gas by the radiation. The roentgen expresses a radiation quantity in terms of air rather than in terms of radiation absorbed by the human body. The unit which has been adoped for radiation absorbed by the body is the *sievert* (Sv) which is defined as an energy absorption of 1 joule per kilogramme. Formerly the unit used was the *rad* (radiation absorbed dose) − 1 Sv = 100 rads. The extent to which a gas is ionised by radiation can be determined by measuring the electrical conductivity of the gas and this principle is used in instruments for the measurement of radioactivity. For all practical purposes the rad, for X-rays and γ-rays with photon energies of less than 3 MeV, can be approximately equated with the roentgen.

Medical authorities consider that there is a maximum permissible radiation dose which can be tolerated by the human body, and this dose is stated as being that amount of radiation which, *in the light of current knowledge*, will not cause appreciable harm to the body over a number of years. The currently accepted dose for classified workers, namely those who are engaged in radiography, is 1 mSv (0.1 rad) for a normal five-day working week. The maximum dosage rate for a year is 50 mSv (5 rads) and the total cumulative dose received by a classified worker should not exceed $(180 + 50N)$Sv, where N is the number of years by which the worker's age exceeds 18. No person aged less than eighteen should be engaged in radiography. It is also considered that persons, other than classified workers, who might work in the general vicinity of radiography activity should not receive more than 15 mSv (1.5 rads) per year and this means that adequate shielding and protection should be provided around X-ray and γ-ray installations.

6.22 Protection against radiation

The intensity of radiation falls off as the square of the distance from the source, and so distance from the source can be an effective and relatively cheap form of protection. However, a fixed X-ray unit is usually housed in a laboratory and the walls of the laboratory are constructed in such a manner as to afford the necessary shielding. The United Kingdom regulations state that radiation on the outer side of the shielding should not exceed 7.5 μSv (0.75 millirads) per hour or, if only classified radiography workers have access to the area, should not exceed 25 μSv (2.5 millirads) per hour. The walls of an X-ray unit, therefore, are usually lined with a thickness of lead or made to have a high absorption factor using barium concrete. Any glazing will be of thick lead-silicate glass. The X-ray unit controls should be placed outside the shielded room.

There are many cases where the material to be radiographed is too large to be taken into an X-ray laboratory and radiography must be carried out on site, for example, in a workshop or aircraft hanger. In these cases it is distance which will give the necessary protection and a sufficiently large area must be roped off and warning signs posted to keep all personnel outside the danger area. It may be possible for the control panel to be placed sufficiently far away from the X-ray or γ-ray source as to be outside the roped-off area. If this is not possible and the control unit has to be close to the source, then the operator must be protected by means of portable lead screening while exposures are made. The criterion is always that a classified worker must not be exposed to radiation at a level in excess of 25 μSv (2.5 millirads) per hour and that the maximum radiation level for any other person must not exceed 7.5 μSv (0.75 millirads) per hour.

6.23 Measurement of radiation received by personnel

The extent of radiation which may be received by classified workers in the field of radiography must be monitored and this is best achieved by recording the dosage received on a radiation monitoring film (film badge) or by using a pocket ionisation chamber. The film badge type of radiation dosemeter is based on the principle that the density recorded on the film is directly related to the amount of radiation to which the film has been exposed. This type of dosemeter consists of a small piece of film packed in a light-tight paper envelope and held within a small plastic container which is pinned or clipped to the operator's outer clothing. The film badge is carried by the operator for some pre-determined time and is then processed under standardised conditions. The density of the processed film is compared with pieces of film of the same type and batch which have been exposed to known levels of radiation and processed under the same conditions.

Pocket-type ionisation dosemeters are usually of similar size to a pen and are carried in the operator's pocket. They possess a scale and pointer which will indicate the dose received in milli-roentgens. It is a statutory requirement in the United Kingdom that full records be kept of the radiation doses received by classified radiographic workers.

7

Other Non-destructive Inspection Techniques

7.1 Optical inspection probes

Optical inspection probes are a major aid to visual inspection as they permit the operator to see clearly inside pipes, ducts, cavities and other openings to which there is limited access. The basic parts of an inspection probe system are the objective lens head which is inserted into the cavity, the viewing eyepiece, and the illumination system. The development of fibre optical systems has permitted major advances to be made in the design and construction of inspection probes.

Optical inspection probes are of two general types, rigid or flexible, but within both of these categories there are many different sizes and designs available.

A rigid inspection probe comprises an optical system with a viewing eyepiece at one end. Illumination is conveyed to the inspection point through an optical fibre bundle and both the optical and illumination systems are enclosed within a stainless steel tube (see figure 7.1).

Light from an external source, which is usually a variable intensity mains and/or battery-operated quartz–halogen lamp, is conveyed to the probe through an optical fibre light guide.

Rigid probes are produced in many sizes from the smallest, with tube diameters of 2 mm or less, up to large probes with tube diameters of 15 or 20 mm. The maximum usable or working length of a probe is the extent to which it can be inserted into an opening; it is not a constant and it varies with the value of probe diameter. Probes of all diameters are made in a variety of lengths but the maximum working length of a 2 mm diameter instrument is about 150 mm. The maximum working length for an 8 mm diameter probe can be up to 2 m and for larger sized probes usable lengths may extend up to 4 or 5 m. It is not practical, even for large diameter devices, to increase the usable length beyond about 5 m, without incurring considerable loss of quality in the eyepiece image.

Inspection probes may be designed to give either direct viewing or to view at some angle to the line of the probe. Some of the viewing angles catered for by the instrument manufacturers are 15°, 60°, 80°, 90° and 120°. In addition to a range

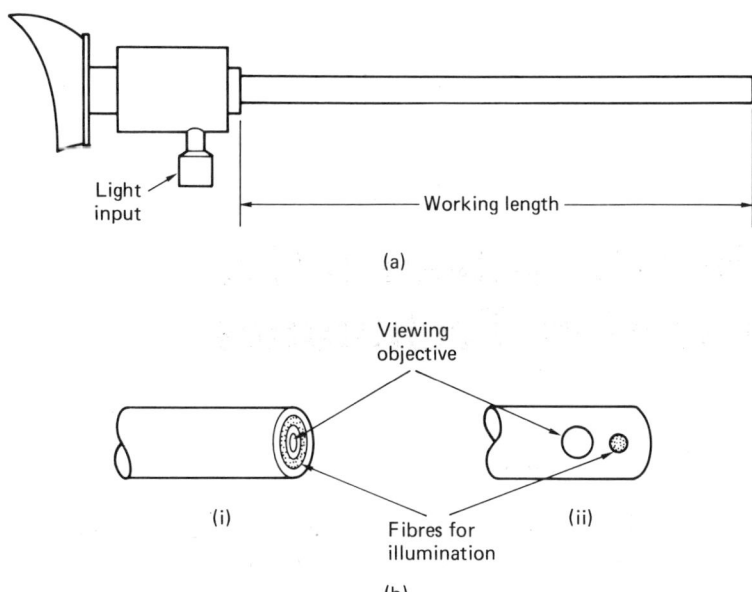

FIGURE 7.1   (a) Rigid optical inspection probe. (b) Probe ends: (i) for direct viewing; (ii) for angled viewing.

of viewing angles the objective lens system can be designed to give a narrow, intermediate or wide field of view. It is also possible to have instruments which possess adjustable prisms to vary the viewing angle. This is only possible in probes with diameters of about 8 mm or greater, but such probes are often made to give any viewing angle between $60°$ and $120°$.

Rigid inspection probes are extremely useful instruments but, like all delicate instruments, have to be handled carefully. The probes, particularly those of small diameter, can be damaged very easily if mis-handled.

The usefulness and versatility of light inspection probes is increased by the use of flexible probes. These incorporate a fibre optic coherent image guide and a separate fibre light guide for illumination, both contained within a flexible plastic or braided metal sheathing. The external diameters of the flexible probes generally range from 4 mm to about 15 mm and the working lengths may be up to about 3 mm. Flexible probes are usually designed to provide either direct viewing ahead or viewing at $90°$ to the probe axis, but the larger diameter probes can be produced with a movable inspection head, the position of which is controlled from the eyepiece end.

Inspection probes, then, are extensions to the human eye, and are used to view areas which would otherwise be impossible to inspect without either dismantling or even cutting open the part or assembly. The images produced at the eyepiece end of a probe may be photographed and a permanent record secured. It is also possible to mount a TV camera as a substitute for the normal eyepiece lens and display the resulting image on a monitor screen. Such an installation is shown in use in figure 7.2. A flexible inspection probe has been inserted into the heart of an aircraft engine. The TV camera is the cylindrical object being held by the operator and the image is visible on the monitor screen. Miniature TV cameras with diameters of the order of 35 mm are now available, and it is possible to insert a probe with camera attached into pipes and ducts, thus effectively increasing the working length of an inspection probe.

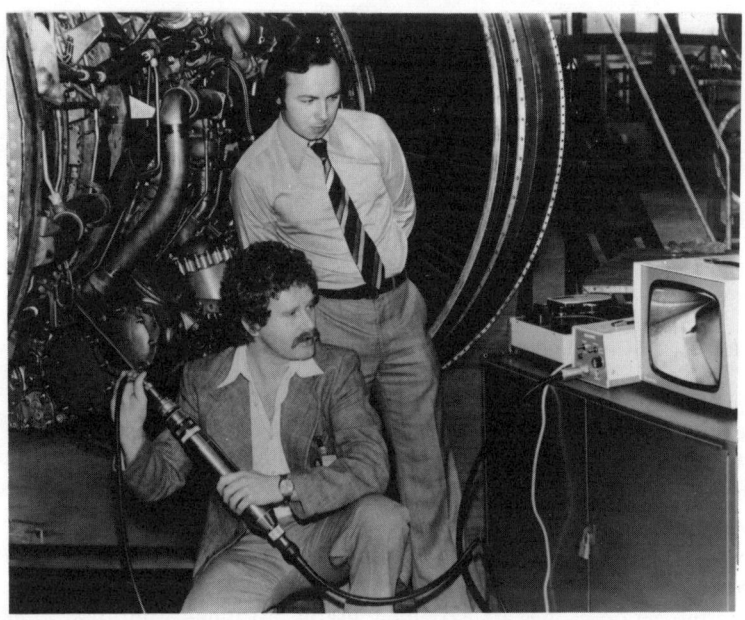

FIGURE 7.2   Inspection probe system with TV camera and monitor (courtesy Inspection Instruments Ltd).

7.2 Neutron radiography

The process of neutron radiography involves the transmission of neutrons through a component or assembly and the production of a radiograph on film. Neutrons are sub-atomic particles which possess a relatively large mass but which carry no electrical charge. Neutron emissions from any source are classified according to their kinetic energies, that is, the speeds of the neutrons in the beam. The types of neutron emissions are termed *cold* (slow neutrons), *thermal* (medium-speed neutrons) or *fast*. In general, it is thermal neutrons with energies in the range 0.01 to 0.3 eV which are used for neutron radiography. A neutron beam passing through a medium will be attenuated but the mechanism of attenuation differs from that for electromagnetic radiation in the X-ray (or γ-ray) sector of the spectrum. Neutron beam attenuation is caused by two effects, scatter and capture.

Neutron scatter Neutrons do not interact with the orbital electrons of atoms but only with atomic nuclei. A neutron may collide with the nucleus of an atom and be deflected. It will transfer some of its kinetic energy to the nucleus at collision and will move away in a new direction with reduced speed.

Neutron capture Some collisions between a neutron and the nucleus of an atom may result in the absorption of the neutron into the nucleus, creating an isotope of the atom with an increased mass number. In some instances the isotope produced by neutron capture may be unstable and undergo radio-active decay.

The attenuation of X-rays or γ-rays when passing through a medium is a direct function of both the density and the atomic mass number of the elements making up the medium, with greater attenuation occurring in the more dense materials and those elements of high mass number. This is not the case for neutron beam attenuation. There is a general tendency for the attenuation of a beam of thermal neutrons to increase as the atomic mass number of the element increases, but it is not a linear function. Certain elements, for example, hydrogen, hafnium, lithium, boron and cadmium, have a much greater attenuation effect on thermal neutrons than other atomic species with similar mass numbers. Also, some specific isotopes of elements show a much greater attenuation than others. Thus, it becomes possible using neutron radiography to detect with ease some light or low atomic number elements even when they are present in assemblies with high atomic number elements such as iron or lead. The degree of discrimination possible in these instances is much greater than would be possible using X-radiography or γ-radiography. The source of a neutron beam may be a radio-active isotope, a particle accelerator or a nuclear reactor. There are several radio-active isotopes which emit neutrons, including antimony-124, polonium-210, and americium-241. In general, the use of neutrons from an isotope source gives radiographs with a poorer resolution than is possible using other neutron sources and exposure times tend to be lengthy. Several types of accelerator have been used for the generation of thermal neutron beams and good resolution can be achieved using these sources. The highest intensity neutron beams can be obtained from nuclear reactors, and radiographs showing very good resolution and requiring short exposure times can be achieved. All the source types emit both thermal and fast neutrons and must be surrounded by a moderator to reduce the energies of the fast neutrons. The neutrons leave the moderator in all directions and, as it is not possible to focus a neutron beam, the emission from the moderator is collimated to produce, in effect, a beam from a point source.

The general principles of beam attenuation with increasing thickness, shadow formation and other geometrical effects are similar to those for X-radiography and γ-radiography. The principles for recording an image in neutron radiography, however, differ somewhat. Radiographic film emulsions are not sensitive to neutrons in the way that they are to electro-magnetic radiation and screens must be used. There are two basic methods, namely direct exposure using screens and the use of a transfer screen.

In the direct exposure method the screen is placed in direct contact with the film. Neutrons falling on the screen are absorbed and a secondary emission, to which the film emulsion is sensitive, is released. One type of screen used is a thin foil of gadolinium in which neutrons are absorbed and β-particles and γ-rays emitted, the intensity of the secondary emission being directly proportional to the incident neutron intensity. Another type of screen used is a zinc sulphide screen impregnated with a small quantity of the lithium-6 isotope. Absorption of neutrons by the lithium atoms occurs, with immediate release of α particles. The α-particles, in turn, cause the zinc sulphide to fluoresce, so exposing the film.

Transfer screens are thin sheets of either indium or dysprosium. The transfer screen alone is positioned behind the object to be radiographed and a neutron exposure made. Neutrons are absorbed, creating an active isotope of either indium or dysprosium which then decays with a short half-life period, emitting β-particles and γ-radiation. After exposure to the neutron beam the transfer screen is placed in contact with a film. The film, thus exposed, may be processed and viewed in the usual way. The transfer screen technique can be used to advantage for the radiography of radio-active materials. γ-radiation emitted from the material would cause film fogging and would make it impracticable to make a radiograph using either X-ray or direct exposure neutron radiography techniques.

Neutron radiography is a highly specialised and costly process, but it has been used for a number of applications including inspection of nuclear reactor fuel elements and assemblies involving both low and high density components such as metal-jacketed explosives and adhesive-bonded metals.

7.3 Laser-induced ultrasonics

A recent development in non-destructive testing is the generation of ultrasonic pulses within a material without the necessity to have a transducer crystal in contact with the testpiece. The ultrasonic pulses are produced by focusing a series of light impulses from a laser on the surface of the testpiece.

The laser, which may be situated up to 10 metres from the testpiece, sends out a series of very short high-energy light impulses and these impulses, each of about 20 nanosecond length, are converted by thermo-mechanical effects into sound impulses at a frequency within the range 1 MHz to 100 MHz. A laser pulse, incident on a solid surface, produces rapid heating of the surface, at the point of incidence, resulting in a localised temperature rise. Thermal expansion of the 'hot zone' causes generation of ultrasonic waves, which propagate across the component surface and within the component body. The intensity of the incident laser impulses is such that no damage is caused to the surface of the testpiece. The emission from a second laser illuminates the surface of the testpiece and ultrasonic echoes returning to the testpiece surface cause deflections which cause a modulation of the reflected light from the illuminating laser (see figure 7.3).

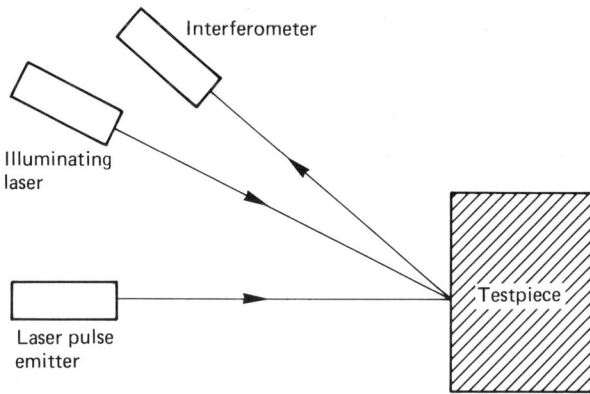

FIGURE 7.3   Laser-stimulated ultrasonic test system.

The third component of the system is an interferometer which analyses the modulated reflected light signal and converts it into a signal which can be presented on the screen of a cathode ray tube in a manner similar to the usual type of ultrasonic signal display.

The main advantages of this technique are that no mechanical coupling is necessary and the acquisition of results is rapid. Laser-based ultrasonic interrogation systems are in use, currently, to detect the existence of piping and liquid metal level in cast steel ingots. Although the sensitivity of the system is lower than that of some of the more conventional techniques, for example, ultrasonic pulse-echo testing, the system has attracted some interest for the continuous monitoring of components on process lines in the manufacturing industries.

7.4 Time-of-flight diffraction

A new ultrasonic technique has been developed, namely time-of-flight diffraction (TOFD), which relies on the diffraction of ultrasonic waves from crack or defect tips, rather than reflection, as in pulse-echo. The technique is very useful in determining the true size of fatigue cracks, even though a crack may be pressed together by the applied load or residual stress network. With conventional pulse-echo testing, complete or partial transmission of the wave pulses across the 'closed' crack can lead to errors in the analysis of crack size, because of the reduction in amplitude of the reflected signals. TOFD is so called because it relies on the wave propagation times to indicate and locate the diffraction source. An example of applying the technique is shown in figure 7.4.

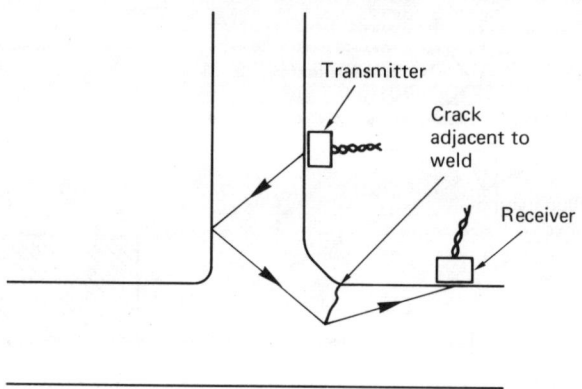

FIGURE 7.4   Probe and wave path geometry as used to measure the size of a crack in a welded joint.

The low signal-to-noise ratio often necessitates signal averaging, and comparison and subtraction of surface waves also may be necessary.

7.5 Acoustic emission inspection

High-frequency waves, at frequencies within the range 50 kHz to 10 MHz, are emitted when strain energy is rapidly released as a consequence of structural changes taking place within a material. Plastic deformation, phase transformations, twinning, micro-yielding and crack growth result in the generation of 'acoustic' signals which can be detected and analysed. Hence, it is possible to obtain information on the location and structural significance of such phenomena.

Basically there are two types of acoustic emission from materials — a continuous type and an intermittent or burst type. Continuous emission is normally of low amplitude and is associated with plastic deformation and the movement of dislocations within a material, while burst emissions are high-amplitude short-duration pulses resulting from the development and growth of cracks.

Acoustic emission inspection offers several advantages over conventional non-destructive testing techniques. For example, it can assess the dynamic response of a flaw to imposed stresses. When a crack or discontinuity approaches critical size there is a marked increase in emission intensity, and hence, a warning is given of instability and catastrophic failure. Also, it is possible to detect growing cracks of about 2×10^{-4} mm in length. This is a much smaller size than is detectable by conventional techniques. In addition, acoustic emission inspection requires only limited access and may be performed directly on components in service.

The usual type of transducer employed for the sensing of acoustic emissions from components uses a lead zirconium titanate element with a high electro-mechanical coupling coefficient. The signal from the transducer is amplified, filtered and processed to give an audio and/or visual recording. A typical arrangement for detecting acoustic emission signals is shown schematically in figure 7.5.

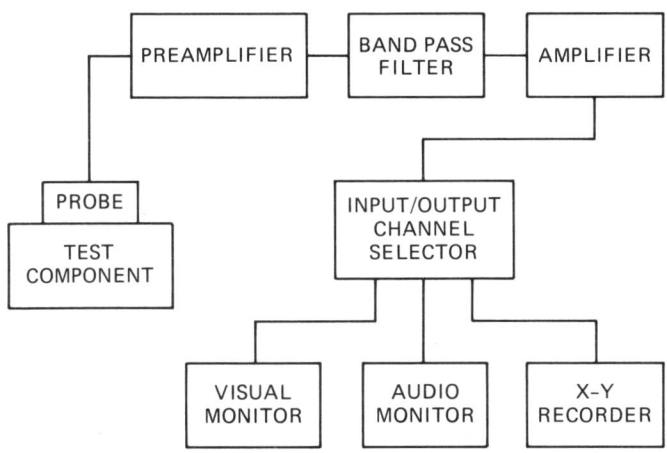

FIGURE 7.5   Schematic diagram for an AE system.

In some applications, for example, inspection of pressure vessels in nuclear plant, it is necessary to detect both the source and location of acoustic emission signals. Several transducers are used, spaced over the surface of the vessel, and computer-assisted monitoring of the time of arrival of signals to the various locations allows analysis of both the source type and its location.

A wide variety of materials can be inspected, including metals, ceramics, polymers, composites and wood. Although the emission sources from each material type may differ, characteristic signals are produced which can be correlated to material integrity. Figure 7.6 shows an acoustic emission trace for a low alloy steel bar subjected to fatigue loading with a mean tensile stress of 500 MN/m^2 at room temperature. During the first 10^5 cycles, the relatively low-amplitude output is an indication of the occurrence of plastic deformation and slip. Above 10^5 cycles, the higher-amplitude output results from the initiation and growth of micro-cracks — that is, the inherent plastic flow at the tips of growing cracks. It can be observed that a clear significant indication or warning of impending fatigue

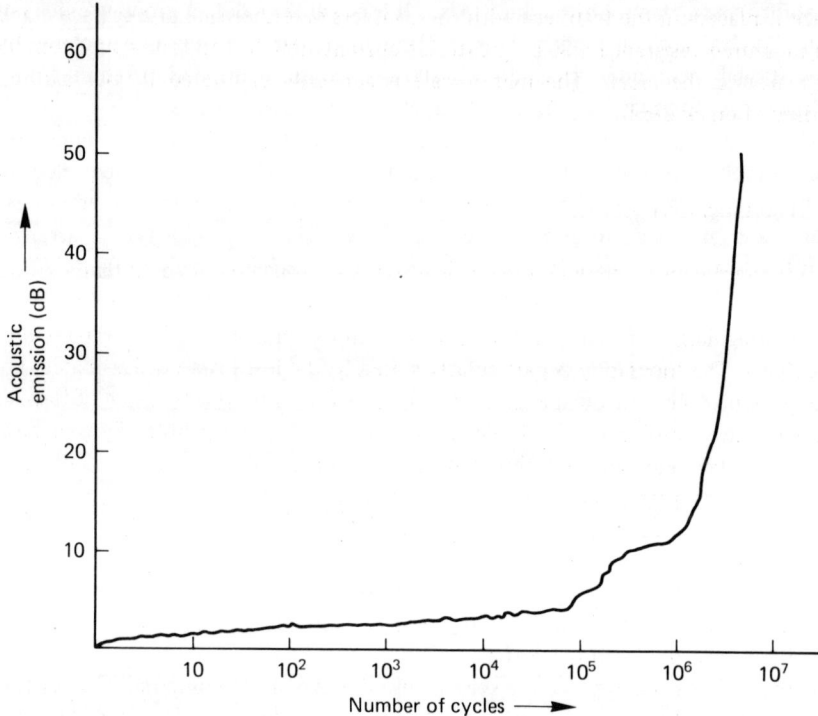

FIGURE 7.6   Acoustic emission trace from a low alloy steel during fatigue.

failure occurs. Acoustic emission inspection can also be employed to monitor stress corrosion and corrosion fatigue failures. One interesting application is the recording of welding 'signatures', since the 'noise' of a welding process is a sum of all associated phase transformations, thermal expansion and contraction, and the formation of cracks and cavities within the welded zone.

7.6 Crack depth gauges

As mentioned in chapters 2 and 3, cracks which appear at the surface of a material can be readily detected using liquid penetrant or magnetic particle inspection methods, but neither of these methods will give an accurate assessment of the depth of a crack. Crack depth gauges are frequently used in conjunction with these other non-destructive tests to give a measure of the depth of flaws which have been located. One simple but effective device for this consists of two closely spaced electrical contacts. The gauge is placed on the surface of the material and the electrical resistance between the two contact points measured. When the gauge is placed on the testpiece with the contacts on either side of a surface crack, the measured resistance will be greater as current now has to follow an extended path around the crack. The meter scale is generally calibrated to give a direct reading of crack depth.

7.7 Thermography

Thermography means the mapping of isotherms, or contours of equal temperature, over the surface of a component. Heat-sensing materials or devices can be used to detect irregularities in temperature contours and such irregularities can be related to defects. Thermography is particularly suited to the inspection of laminates. The conduction of heat through a laminate will be affected by the presence of flaws in the structure, resulting in an irregular surface temperature profile. Typical flaws which can be detected are unbonded areas, crushed cells, separation of the core from the face plates and the presence of moisture in the cells of honeycomb structures.

Thermographic methods may either be of the direct contact type, in which heat-sensitive material is in contact with the component surface, or indirect contact, in which a heat-sensitive device is used to measure the intensity of infrared energy emitted from the surface.

Pulses of heat energy, from a source, are directed at the component under test. It is usual, but not essential, to direct the incident energy on to one surface of a component and observe the effects at the opposite surface after conduction through the material. Flaws and irregularities in structure will affect the amount of conduction in their vicinity. If it is impossible to have access to both surfaces, the technique can still be used. The heat energy incident on the surface will be conducted away through the material at differing rates, depending on whether or not flaws are present.

Direct contact methods include the use of heat-sensitive paints and thermally quenched phosphors. Indirect contact methods, which offer greater sensitivity, involve the use of infra-red imaging systems with a TV-video output.

Heat-sensitive paints

Heat-sensitive photo-chromic paints are effective over a temperature range from about 40°C to 160°C. Some paints show several colour changes within their reaction temperature range and, with careful application, will have a sensitivity of the order of ±5°C. When heat reaches the painted surface by conduction through the material the paint colour changes, usually with a bleaching effect. Where a flaw impedes conduction the colour will be unchanged. On the other hand, if heat energy is directed at the painted surface the reverse effect will show up as heat is conducted away from the surface more rapidly through good regions than through defective areas.

Thermally quenched phosphors

These are organic compounds which emit visible light when excited by ultra-violet radiation. The brightness of the emission decreases as the temperature of the compounds increases. Phosphors are available that are useful at temperatures up to about $400°C$ and with a resolution of $±1°C$.

Thermal pulse video thermography

In this system no physical contact is necessary with the material and very rapid rates of inspection are possible. A high-intensity heating source is used to send pulses of infra-red energy into the material. The surface is scanned by an infra-red thermal imager with a TV-video output. This system, again, can either be used for sensing heat transmitted through the component or for single-sided inspection when only one surface is accessible. Very good sensitivities are possible. Digitised image processing to provide image enhancement is also possible.

7.8 Surface texture analysis

A recent development of particular interest is a capacitance-based system for the rapid measurement of the surface finish or texture of engineering components. This parameter is of importance to the function and service life of components, such as engine valves, compressor wheels, turbine blades, bearing cages and extrusion dies.

The 'roughness' of a surface is a function of the 'air space' between the peaks and valleys of the surface. Air is an electrical insulator and can act as the dielectric of a capacitor. The electrical capacitance of a capacitor, C, changes as the amount or average height, t, of dielectric between the plates is changed. If a surface to be measured is made, temporarily, into a capacitor by placing an electrode plate on the surface, the air space acts as the capacitor dielectric. Hence, the resultant capacitance depends on the roughness of the surface finish. Figure 7.7 illustrates the principle of operation. Provided that the probe has been calibrated on reference specimens of known surface roughness assessed by stylus measurement, the system permits accurate determination of surface roughness, in either micro-inches or microns. The result is displayed as a digital read-out on the ancillary instrumentation.

7.9 Multi-phase flow analysis

A major parameter in the design and use of safe, efficient and economic oil and gas pipelines is the prediction of multi-phase flow behaviour of the transferred medium. A multi-path ultrasonic gas flowmeter has been designed and developed by British Gas to meet this requirement. The equipment has a variety of applica-

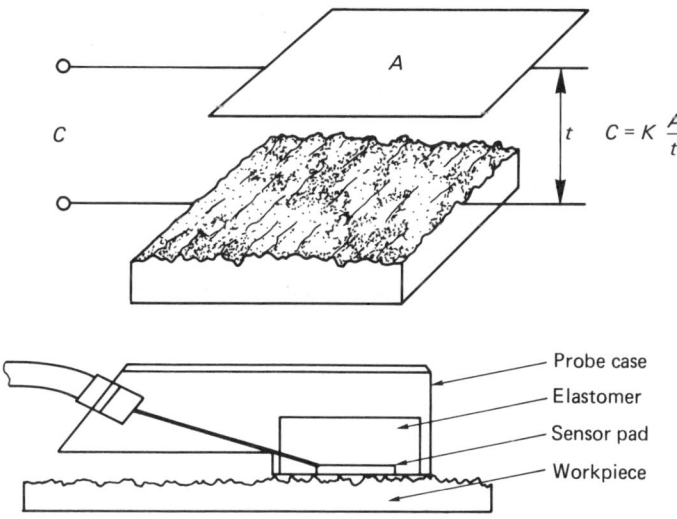

FIGURE 7.7   Principle of operation of capacitance probe.

tion areas, such as shore terminals, transfer and compressor stations, and underground storage sites. The principle of operation of the flowmeter is based on ultrasonic pulse transit time within the flowing medium. A simple twin transducer set up is shown in figure 7.8.

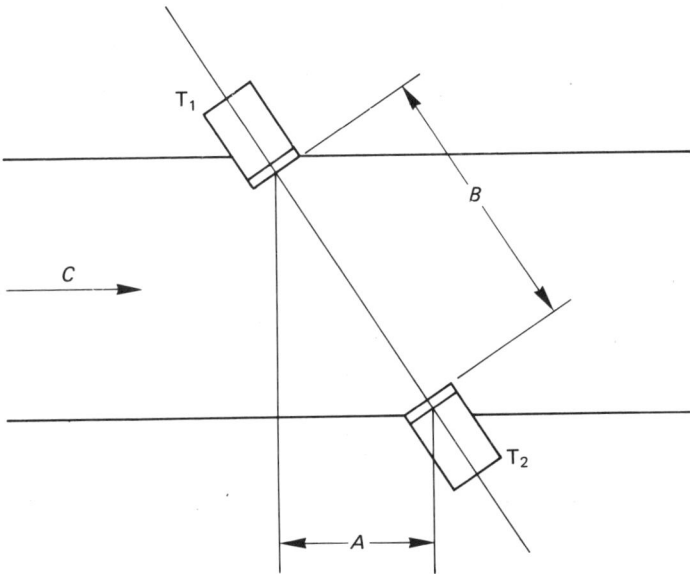

FIGURE 7.8   Schematic diagram of an ultrasonic flowmeter in operation.

The transit time, t, for ultrasonic pulses to travel between a transmitter and receiver is measured with a high-frequency, electronic clock. A pulse travelling from T_1 to T_2 is increased in velocity by the resolved component of the gas/oil flow along its path. Hence, its mean velocity is $(V + c(A/B))$ where V is the velocity of sound and c is the mean flow velocity of the medium. A and B are the dimensions as shown in figure 7.8. The time taken for a pulse to travel from T_1 to T_2, t_1, is given by

$$t_1 = \frac{B}{V + c(A/B)}$$

or

$$\frac{1}{t_1} = \frac{V + c(A/B)}{B}$$

Similarly, a pulse travelling in the opposite direction, that is, from T_2 to T_1, is slowed by the resolved component of the medium flow rate and takes time t_2. Hence

$$\frac{1}{t_2} = \frac{V - c(A/B)}{B}$$

Subtracting

$$\frac{1}{t_1} - \frac{1}{t_2} = \frac{2cA}{B^2}$$

therefore

$$\frac{t_2 - t_1}{t_1 t_2} = \frac{c}{k_1}$$

where $k_1 = \dfrac{B^2}{2A}$ (a constant), or

$$c = \frac{k_1 (t_2 - t_1)}{t_1 t_2}$$

The volume flow rate $Q = cx$, where x is the cross-sectional area of the pipe. Therefore

$$Q = \frac{k_2 (t_2 - t_1)}{t_1 t_2}$$

where $k_2 = k_1 x$ (a constant).

A complete flowmeter system comprises four pairs of transducers, eight transducers in total, which are aligned to determine the flow rates in the upper, central and lower parts of a pipeline. In addition, the system accepts temperature, pressure, density and calorific value inputs, and hence volume flowrates, mass flow rates and heat energy flow rates can be obtained.

7.10 Conclusions

The techniques mentioned in sections 7.2 to 7.8 above are a selection of some of the more recent developments in non-destructive inspection. It will be apparent that within the realm of inspection there is continual progress, with existing techniques being improved and refined and their range of applications extended, and new processes being created to meet increasingly demanding inspection requirements.

It would probably be true to say that wherever an inspection problem exists a solution can be found. The solution may be obtained by adapting some existing technique or, in the longer term, by developing a new inspection process.

Bibliography

Books and Handbooks

ASM Metals Handbook, Vol.11, *Non-destructive Inspection and Quality Control*, American Society of Metals.

J. C. Drury, *Ultrasonic Flaw Detection for Technicians*, The Unit Inspection Co. Ltd.

R. Halmshaw (Editor), *Mathematics and Formulae in NDT*, British Institute of Non-destructive Testing.

Handbook of Radiographic Apparatus and Techniques, International Institute of Welding.

Handbook on the Ultrasonic Examination of Welds, International Institute of Welding.

Industrial Radiography, Agfa-Gevaert Ltd.

Industrial Radiography, Kodak Ltd.

Krautkramer Booklet, Krautkramer GMBH, Cologne.

M. G. Silk, *Ultrasonic Transducers for Non-destructive Testing*, Adam Hilger Ltd.

J. L. Taylor, *Basic Metallurgy for Non-destructive Testing*, British Institute of Non-destructive Testing.

Ultrasonic Non-destructive Testing, Baugh and Weedon Ltd.

British Standards

BS 2600, *Radiographic examination of fusion welded butt joints in steel.*

BS 2704, *Specification for calibration blocks for use in ultrasonic flaw detection.*

BS 2737, *Terminology of internal defects in castings as revealed by radiography.*

BS 2910, *Method for radiographic examination of fusion welded circumferential butt joints in steel pipes.*

BS 3683 Parts 1–5, *Glossary of terms used in non-destructive testing.*

BS 3889, *Methods for non-destructive testing of pipes and tubes.*

BS 3923, *Methods for ultrasonic examination of welds.*

BS 3971, *Specification for image quality indicators for industrial radiography.*

BS 4069, *Specification for magnetic flaw detection inks and powders.*

BS 4094, *Recommendation for data on shielding from ionising radiation.*

BS 4124, Part 1, *Methods for non-destructive testing of steel forgings – Ultrasonic flaw detection.*

BS 4331, *Methods for assessing the performance characteristics of ultrasonic flaw detection equipment.*

BS 4408, Part 3, *Gamma radiography of concrete.*

BS 4489, *Method for measurement of UV-A radiation (black light) used in non-destructive testing.*

BS 5044, *Specification for contrast aid paints used in magnetic particle flaw detection.*

BS 5138. *Specification for magnetic particle flaw inspection of finished machined solid forged and drop stamped crankshafts.*

BS 5411, Parts 1, 2, 3 & 11, *Methods of test for metallic and related coatings.*

BS 5650, *Specification for apparatus for gamma radiography.*

BS 5996, *Methods for ultrasonic testing and specifying quality grades of ferritic steel plates.*

BS 6072, *Method for magnetic particle flaw detection.*

BS 6208, *Methods for ultrasonic testing and for specifying quality levels of ferritic steel castings.*

BS 6443, *Method for penetrant flaw detection.*

PD 6513, *Magnetic particle flaw detection – A guide to the principles and practice of applying magnetic particle flaw detection.*

Aero M34, *Method of preparation and use of radiographic techniques.*

Aero M36, *Method for ultrasonic testing of special forgings by an immersion technique.*

Aero M39, *Method for penetrant inspection of aerospace products.*

Aero M40, *Methods for measuring coating thickness by non-destructive testing.*

Aero M42, *Methods for non-destructive testing of fusion and resistance welds in thin gauge material.*

Note: All the above standards publications are available from the British Standards Institution, Linford Wood, Milton Keynes, MK14 6LE.

Index

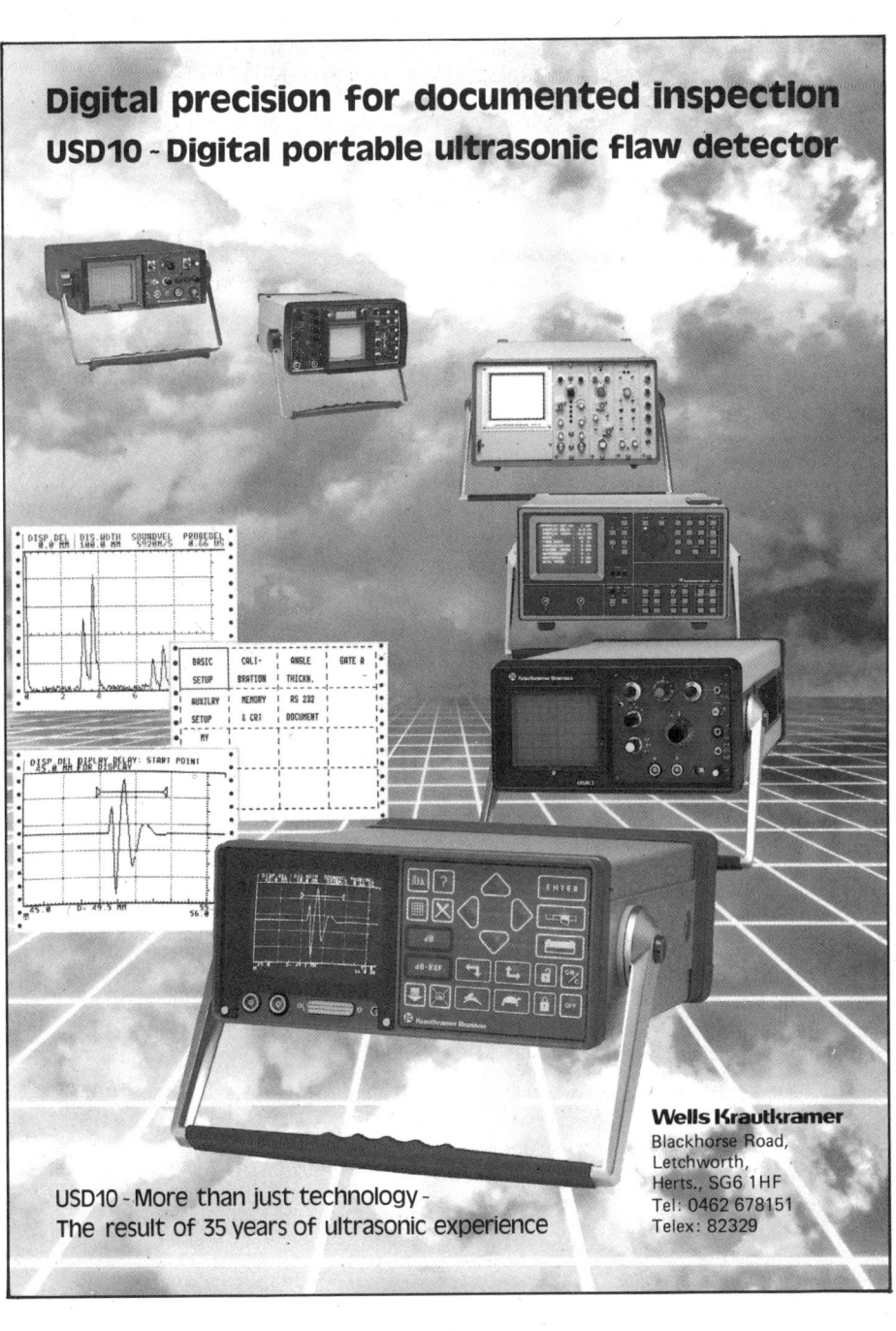